Mathematical Optimization
in Computer Graphics and Vision

THE MORGAN KAUFMANN SERIES IN COMPUTER GRAPHICS

Mathematical Optimization in Computer Graphics and Vision

PAULO CEZAR PINTO CARVALHO

LUIZ HENRIQUE DE FIGUEIREDO

JONAS GOMES

LUIZ VELHO

AMSTERDAM • BOSTON • HEIDELBERG • LONDON
NEW YORK • OXFORD • PARIS • SAN DIEGO
SAN FRANCISCO • SINGAPORE • SYDNEY • TOKYO

Morgan Kaufmann Publishers is an imprint of Elsevier

MORGAN KAUFMANN PUBLISHERS

Publishing Director: Chris Williams
Senior Acquisitions Editor: Tiffany Gasbarrini
Publishing Services Manager: George Morrison
Production Editor: Mageswaran BabuSivakumar, Lianne Hong
Assistant Editor: Matthew Cater
Cover Design: Dennis Schaefer
Cover Photo Image: © Masterfile (Royalty-Free Division)
Composition: diacriTech
Copyeditor: diacriTech
Proofreader: diacriTech
Indexer: diacriTech
Interior printer: Sheridan Books, Inc.
Cover printer: Phoenix Color Corporation

Morgan Kaufmann Publishers is an imprint of Elsevier.
30 Corporate Drive, Suite 400, Burlington, MA 01803, USA

Library of Congress Cataloging-in-Publication Data
Application submitted

ISBN: 978-0-12-715951-5

For information on all Morgan Kaufmann publications,
visit our Web site at *www.mkp.com* or *http://www.books.elsevier.com*

Printed and bound by CPI Group (UK) Ltd, Croydon, CR0 4YY

Transferred to Digital Print 2012

Contents

List of Figures

Chapter 3

Chapter 4

Chapter 5

Chapter 6

Chapter 7

Chapter 8

Preface

Mathematical optimization has a fundamental importance in solving many problems in computer graphics and vision. This fact is apparent from a quick look at the SIGGRAPH proceedings and other relevant publications in these areas, where a significant percentage of the papers use mathematical optimization techniques.

The book provides a conceptual analysis of the problems in computer graphics and discusses the mathematical models used to solve these problems. This motivates the reader to understand the importance of optimization techniques in graphics and vision.

The book gives an overview of combinatorial, continuous, and variational optimization methods, focusing on graphical applications. The prerequisites for the book are (i) background on linear algebra and calculus of one and several variables; (ii) computational background on algorithms and programming; and (iii) knowledge of geometric modeling, animation, image processing, image analysis, and visualization.

This book originated from a set of notes in Portuguese that we wrote for a course on this topic at the Brazilian Mathematical Colloquium in July 1999 and at Brazilian Congress of Applied Mathematics in October 2000. After that, these notes were expanded and translated to English. This material was then presented in two highly successful SIGGRAPH tutorial courses in August 2002 and 2003.

Mathematical optimization is vast and has many ramifications, encompassing many disciplines ranging from pure mathematics to computer sciences and engineering. Consequently, there exists a vast literature on this subject, including textbooks, tutorials, and research papers that cover in detail practically every aspect of the field. The sheer amount of information available makes it difficult for a nonspecialist to explore the literature and find the right optimization technique for a specific application.

From the comments above, it becomes apparent that any attempt to present the whole area of mathematical optimization in detail would be a daunting task, probably doomed to failure. This endeavor would be even more difficult if the goal is to understand the applications of optimization methods to a multidisciplinary and diversified area such as computer graphics.

The objective of this book is to give an overview of the different aspects of mathematical optimization to enable the reader to have a global understanding of the field and pursue studies of specific techniques related to applications in graphics and vision. Since mathematical optimization is so pervasive in graphical applications, we decided to present in detail only the seminal techniques that appeared in early papers. The more recent research is discussed in a section at the end of every chapter devoted to comments and references. This section also includes pointers to the literature.

The book is conceptually divided into five parts: computer graphics and optimization; variational and continuous optimization; combinatorial optimization; global optimization methods; and probability and optimization.

The first part of the book gives a conceptual overview of computer graphics, focusing on problems and describing the mathematical models used to solve these problems. This is a short introduction to motivate the study of optimization techniques. The subject is studied in such a way to make clear the necessity of using optimization techniques in the solution to a large family of problems.

The second part of the book discusses the optimization of functions $f: S \to \mathbb{R}$, where S is a subset of the Euclidean space $\mathbb{R}^n$. Optimality conditions are discussed for optimization with and without restrictions. Different algorithms are described such as Newton methods, linear programming, conjugate gradient, sequential quadratic programming. Applications to camera calibration, color correction, animation, and visualization are given. The problem of variational optimization is properly posed, with special emphasis on variational modeling using plane curves.

In the third part of the book, the problem of combinatorial optimization is posed and different strategies to devise good algorithms are discussed. Special emphasis is given to dynamic programming and integer programming. Algorithms of Dijkstra and branch-and-bound are discussed. Examples are given to image and color quantization, level of detail computation, interactive visualization, minimum paths in maps, and surface reconstruction from sections.

In the fourth part of the book, the difficulties of global optimization are clearly pointed and some strategies to find solutions are discussed, including simulated annealing and genetic algorithms. The role of interval and affine arithmetic in the solution to problems is also discussed. Applications to geometric modeling and animation are given.

The fifth part of the book discusses the subtle relationships between probability theory and mathematical optimization. It covers information theory and applications to coding and compression. Notions such as entropy and mutual information are discussed together with applications to image processing and computer vision.

ACKNOWLEDGMENTS

The authors would like to thank the many who assisted with the preparation and review of this book. We also would like to acknowledge the anonymous reviewers who provided helpful critiques of the manuscript. Special thanks goes to Anderson Mayrink who collaborated with us to write the chapter on Probability and Optimization. We appreciate the assistance of Mageswaran BabuSivakumar in the production of the book. Finally, we are very grateful to Diane Cerra and Tiffany Gasbarrini for all their encouragement and editorial support.

1 COMPUTER GRAPHICS

1.1 WHAT IS COMPUTER GRAPHICS?

The usual definition of computer graphics is the following: *a set of models, methods, and techniques to transform data into images that are displayed in a graphics device.* The attempt to define an area is a difficult, if not an impossible, task. Instead of trying to devise a good definition, perhaps the best way to understand an area is through a deep knowledge of its problems and the methods to solve them. From this point of view, the definition above has the virtue of emphasizing a fundamental problem of computer graphics: *the transformation of data into images* (Figure 1.1).

In applied mathematics, the solution to problems is directly related to the mathematical models used to understand and pose the problem. For this reason, the dividing line between *solved* and *open* problems is more subtle than in the case of pure mathematics. In fact, in pure mathematics, different solutions to the same problem, in general, do not constitute great innovations from the scientific point of view; on the other hand, in applied mathematics, different solutions to the same

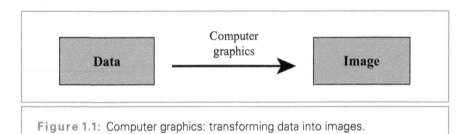

Figure 1.1: Computer graphics: transforming data into images.

problem as a consequence of the use of different models usually bring a significant advance in terms of applications.

This book discusses the solution to various problems in computer graphics using optimization techniques. The underlying idea is to serve as a two-way channel: stimulate the computer graphics community to study optimization methods and call attention of the optimization community to the extremely interesting real-world problems in computer graphics.

1.1.1 RELATED AREAS

Since its origin, computer graphics is concerned with the study of models, methods, and techniques that allow the visualization of information using a computer. Because, in practice, there is no limitation to the origin or nature of the data, computer graphics is used today by researchers and users from many different areas of human activity.

The basic elements that constitute the fundamental problem of computer graphics are *data* and *images*. There are four related areas that deal with these elements, as illustrated in Figure 1.2.

Geometric modeling deals with the problem of describing, structuring, and transforming geometric data in the computer.

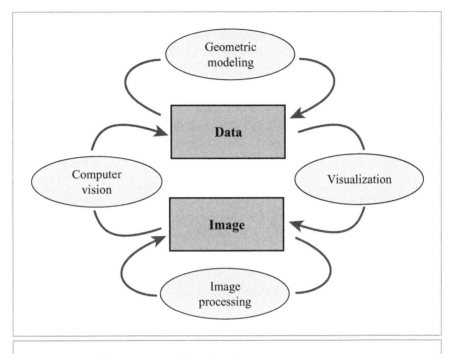

Figure 1.2: Computer graphics: related areas.

Visualization interprets the data created by geometric modeling to generate an image that can be viewed using a graphical output device.

Image processing deals with the problem of describing, structuring, and transforming images in the computer.

Computer vision extracts from an input image various types of information (geometric, topological, physical, etc.) about the objects depicted in the image.

In the literature, computer graphics is identified with the area called visualization (Figure 1.2). We believe it is more convenient to consider computer graphics as the *mother area* that encompasses these four subareas: geometric modeling, visualization, image processing, and computer vision. This is justified because these related areas work with the same objects (data and images), which forces a strong trend of integration. Moreover, the solutions to certain problems require the use of methods from all these areas under a unified framework.

As an example, we can mention the case of a Geographical Information System (GIS) application, where a satellite image is used to obtain the elevation data of a terrain model and a three-dimensional (3D) reconstruction is visualized with texture mapping from different view points.

Combined techniques from these four related subareas comprise a great potential to be exploited to obtain new results and applications of computer graphics. This brings to computer graphics a large spectrum of models and techniques providing a path for new achievements: the combined result is more than the sum of parts.

The trend is so vigorous that new research areas have been created. This was the case of *image-based modeling and rendering*, a recent area that combines techniques from computer graphics, geometric modeling, image processing, and vision. Furthermore, in some application domains such as GIS and medical image, it is natural to use techniques from these related areas to the point that there is no separation between them.

We intend to show in this book how mathematical optimization methods can be used in this unified framework in order to pose and solve a reasonable number of important problems.

1.1.2 IS THERE SOMETHING MISSING?

The diagram given in Figure 1.2 is classical. But it tells us only part of the computer graphics script: the area of geometric modeling and the related areas. In our holistic view of computer graphics, several other areas are missing on the diagram. Where is animation? Where is digital video?

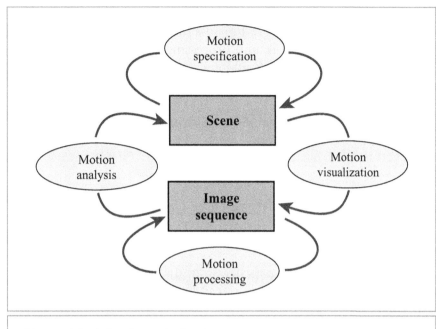

Figure 1.3: Animation and related areas.

In fact, it is possible to reproduce a similar diagram for animation, and this is shown in Figure 1.3.

Note the similarity to the previous diagram. *Motion modeling* is the area that consists in describing the movement of the objects on the scene. This involves time, dynamic scene objects, trajectories, and so on. This is also called *motion specification*. Similarly, we can characterize *motion analysis*, *motion synthesis*, and *motion processing*.

We could now extend this framework to digital video, but this is not the best approach. In fact, this repetition process is an indication that some concept is missing, a concept that could put together all of these distinct frameworks into a single one. This is the concept of *graphical object*.

In fact, once we have devised the proper definition of a graphical object, we could associate with it the four distinct and related areas: modeling, visualization, analysis, and processing of graphical object. The concept of graphical object should encompass, in particular, the geometric models of any kind, animation, and digital video.

We introduce this concept and develop some related properties and results in the next two sections.

1.2 MATHEMATICAL MODELING AND ABSTRACTION PARADIGMS

Applied mathematics deals with the application of mathematical methods and techniques to problems of the real world. When computer simulations are used in the solution to problems, the area is called *computational mathematics*. In order to use mathematics in the solution to real-world problems, it is necessary to devise good mathematical models that can be used to understand and pose the problem correctly. These models consist in abstractions that associate real objects from the physical world with mathematical abstract concepts. Following the strategy of dividing to conquer, we should create a hierarchy of abstractions, and for each level of the hierarchy, we should use the most adequate mathematical models. This hierarchy is called an *abstraction paradigm*.

In computational mathematics, a very useful abstraction paradigm consists of devising four abstraction levels, called *universes*: the physical universe, the mathematical universe, the representation universe, and the implementation universe (Figure 1.4).

The *physical universe* contains the objects from the real (physical) world, which we intend to study; the *mathematical universe* contains the abstract mathematical description of the objects from the physical world; the *representation universe* contains discrete descriptions of the objects from the mathematical universe; and the *implementation*

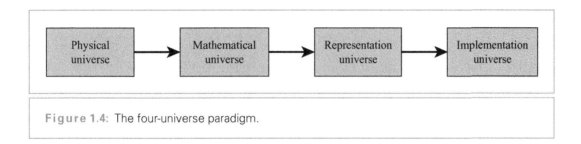

Figure 1.4: The four-universe paradigm.

universe contains data structures and machine models of computation so that objects from the representation universe can be associated with algorithms in order to obtain an implementation in the computer.

Note that by using this paradigm, we associate with each object from the real (physical) world three mathematical models: a continuous model in the mathematical universe, a discrete model in the representation universe, and a finite model in the implementation universe. This abstraction paradigm is called the *four-universe paradigm* (Gomes and Velho, 1995).

In this book, we take the computational mathematics approach to computer graphics. That is, a problem in computer graphics will be solved considering the four-universe paradigm. This amounts to saying that computer graphics is indeed an area of computational mathematics. As a consequence, the four-universe paradigm will be used throughout the book.

Example 1 [length representation]. Consider the problem of measuring the length of objects from the real world. To each object, we should associate a number that represents its length. In order to attain this, we must introduce a standard unit of measure, which will be compared with the object to be measured in order to obtain its length.

From the point of view of the four-universe paradigm, the mathematical universe is the set $\mathbb{R}$ of real numbers. In fact, to each measure, we associate a real number; rational numbers correspond to objects that are commensurable with the adopted unit of measure, and irrational numbers are used to represent the length of objects that are incommensurable.

In order to represent the different length values (real numbers), we must look for a discretization of the set of real numbers. A widely used technique is the floating-point representation. Note that in this representation, the set of real numbers is discretized in a finite set of rational numbers. In particular, this implies that the concept of commensurability is lost when we pass from the mathematical to the representation universe.

From the point of view of the implementation universe, the floating-point representation can be attained by use of the IEEE standard, which is implemented in most of the computers. A good reference for this topic is Higham (1996).

The previous example, although simple, illustrates the fundamental problem of modeling in computational mathematics. Of particular importance in this example is the loss of information we face when moving from the mathematical universe to the representation universe (from real numbers to their floating-point representation), where the concept of commensurability is lost. In general, passing from the mathematical universe to the representation universe implies a loss of information. This is a very subtle problem we must face when working with modeling in computational mathematics and, in particular, in computer graphics.

The modeling process consists in choosing an object from the physical world, associate with it a mathematical model, discretize it, and implement it. This implementation should provide solutions to properly posed problems involving the initial physical object. Note that the loss of information we mentioned above is a critical factor in this chain.

We should remark that by the nature of the modeling process, the relationship between an object defined at one level of abstraction (universe) with the same object defined at another level of abstraction can be the most generic possible. In fact, the same object from one abstraction level may have distinct correspondents in the next abstraction level (we leave to the reader the task of providing examples for this). Instead, in the next two sections, we provide two examples that show that different objects from the physical world may be described by the same mathematical model.

1.2.1 TERRAIN REPRESENTATION

Consider the problem of representing a terrain in the computer (a mountain for instance). In cartography, a terrain may be described using a height map: we devise a reference level (e.g., the sea level), and we take for each point the terrain height at this point. In the mathematical universe, this map of heights corresponds to a function $f: U \subset \mathbb{R}^2 \to \mathbb{R}, z = f(x, y)$, where (x, y) are the coordinates of a point of the plane $\mathbb{R}^2$ and z is the corresponding height. Geometrically, the terrain is described by the graph $G(f)$ of the *height function f*:

$$G(f) = \{(x, y, f(x, y)); (x, y) \in U\}.$$

Figure 1.5 shows a sketch of the graph of the height function of part of the Aboboral mountain in the state of São Paulo, Brazil (Yamamoto, 1998).

How can we represent the terrain? A simple method consists in taking a uniform partition

$$P_x = \{x_0 < x_1 < \cdots < x_n\}$$

of the x-axis,[1] a uniform partition

$$P_y = \{y_0 < y_1 < \cdots < y_m\},$$

---

1   Uniform indicates that $\Delta_j = x_j - x_{j-1}$ is constant for any $j = 1, \ldots, n$.

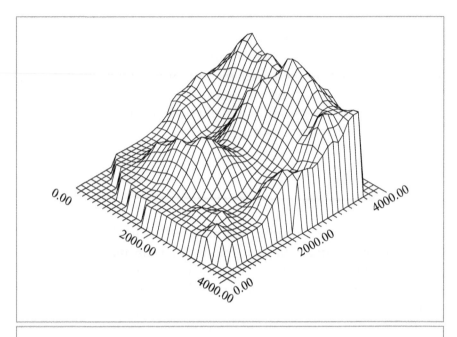

Figure 1.5: Aboboral mountain, scale 1:50000 (Yamamoto, 1998).

of the y-axis, and the Cartesian product of both partitions to obtain
the grid of points

$$(x_i, y_j), i = 1, \ldots, n, \quad j = 1, \ldots, m$$

in the plane as depicted in Figure 1.6. In each vertex (x_i, y_j) of the grid,
we take the value of the function $z_{ij} = f(x_i, y_j)$ and the terrain representation if defined by the height matrix (z_{ij}). This representation
is called *representation by uniform sampling* because we take the vertices of a uniform grid and we sample the function at these points.
The implementation can be easily attained using a matrix as a data
structure.

1.2.2 IMAGE REPRESENTATION

Now we consider the problem of image modeling. We take a photograph as the image model in the real world. This physical image model

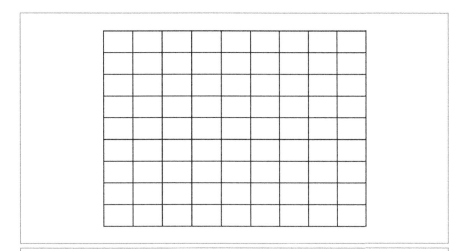

Figure 1.6: A grid on the function domain.

is characterized by two properties:

- It has a support set (a rectangular piece of paper).
- There is a color attribute associated with each point of the support set.

Suppose that we have a black and white photo, indicating that we associate a different luminance of black (gray value) with each point of the support set. This is called a grayscale image in the computer graphics literature. In this case, the "color" of each point can be modeled by a real number in the interval $[0, 1]$, where 0 represents black, 1 represents white, and each number $0 < t < 1$ represents a gray value in between. The rectangular support set of the image can be modeled in the mathematical universe as a rectangular subset $U \subset \mathbb{R}^2$ of the plane.

Therefore, the mathematical model of a grayscale image is a function $f: U \subset \mathbb{R}^2 \to \mathbb{R}$ that maps each point (x, y) to the value $z = f(x, y)$, which gives the corresponding gray value at the point. This function is called the *image function*. Figure 1.7 shows an image on the left and the graph of the corresponding image function on the right.

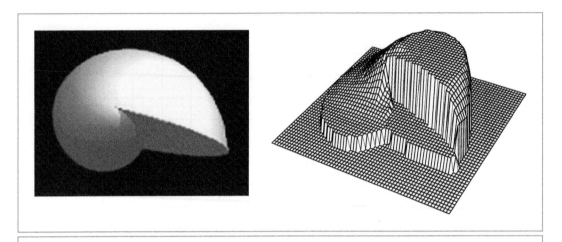

Figure 1.7: Grayscale image and the graph of the image function.

In sum, we see that an image and a terrain can be modeled by the same mathematical object: a function $f: U \subset \mathbb{R}^2 \to \mathbb{R}$. Therefore, we can use uniform sampling to represent the image as we did for the terrain.

The representation of a terrain and an image by a function is not accidental. In fact, a huge number of objects from the physical world are represented by functions $f: X \to Y$, where X and Y are properly chosen topological spaces (in general, they are subsets of Euclidean spaces). On the other hand, in the next section, we characterize a graphical object by a function.

1.3 GRAPHICAL OBJECTS

In the quest for a better definition of computer graphics, we could define *computer graphics as the area that deals with the description, analysis, and processing of graphical objects*. This definition only makes sense if we are able to devise a precise definition of the concept of a graphical object. We do this as follows. A *graphical object* is a subset $S \subset \mathbb{R}^m$ together with a function $f: S \subset \mathbb{R}^m \to \mathbb{R}^n$.

The set S is called *geometric support*, and f is called *attribute function* of the graphical object. The dimension of the geometric support S is called *dimension* of the graphical object. This concept of graphical object was introduced in the literature by Gomes *et al.* (1996).

Let us take a look at some concrete examples to clarify the definition.

Example 2 [subsets of Euclidean space]. Any subset of the Euclidean space $\mathbb{R}^m$ is a graphical object. Indeed, given $S \subset \mathbb{R}^m$, we define immediately an attribute function

$$f(p) = \begin{cases} 1 & \text{if } p \in S, \\ 0 & \text{if } p \notin S. \end{cases}$$

It is clear that $p \in S$ if and only if $f(p) = 1$. In general, the values of $f(p) = 1$ are associated with a color, called *object color*. The attribute function in this case simply characterizes the points of the set S, and for this reason, it is called *characteristic function* of the graphical object.

The characteristic function completely defines the geometric support of the graphical object, that is, *if p is a point of the space $\mathbb{R}^m$, then $p \in S$ if and only if $f(p) = 1$.*

The two problems below are, therefore, equivalent:

1. Devise an algorithm to compute $f(p)$ at any point $p \in \mathbb{R}^m$.
2. Decide if a point $p \in \mathbb{R}^n$ belongs to the geometric support S of the graphical object.

The second problem is called a *point-membership classification problem*. A significant part of the problems in the study of graphical objects reduces to a solution of point-membership classification. Therefore, the existence of robust and efficient algorithms to solve this problem is of great importance.

Example 3 [image]. A grayscale image (see Section 1.2.2) is a function $f: U \subset \mathbb{R}^2 \to \mathbb{R}$, therefore it is a graphical object. More generically, an image is a function $f: U \subset \mathbb{R}^2 \to \mathbb{R}^n$, where $\mathbb{R}^n$ is the representation of a color space. In this way, we see that an image is a graphical object whose geometric support is the subset U of the plane (in general, a rectangle), and the attribute function associates a color with each point of the plane.

Example 4 [circle and vector field]. Consider the unit circle S^1 centered at the origin, whose equation is given by

$$x^2 + y^2 = 1.$$

A mapping of the plane $N: \mathbb{R}^2 \to \mathbb{R}^2$, given by $N(x, y) = (x, y)$, defines a field of unit vectors normal to S^1. The map $T: \mathbb{R}^2 \to \mathbb{R}^2$, given by $T(x, y) = (y, -x)$, defines a field of vectors tangent to the circle (Figure 1.8).

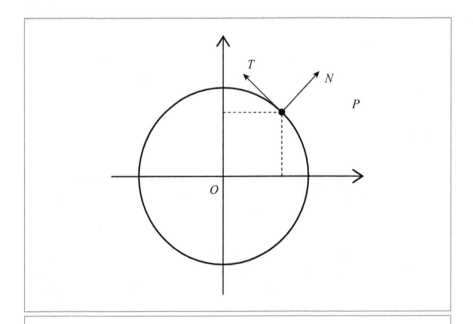

Figure 1.8: Circle with normal and tangent vector fields.

The circle is a 1D graphical object of the plane, and the two vector fields are attributes of the circle (they can represent, for example, physical attributes such as tangential and radial accelerations). The attribute function is given by $f: S^1 \rightarrow \mathbb{R}^4 = \mathbb{R}^2 \oplus \mathbb{R}^2$, $f(p) = (T(p), N(p))$.

1.3.1 REVISITING THE DIAGRAM

Now that we have discussed the concept of graphical object, it is time to go back to Figure 1.2.

Every output graphics device has an internal representation of the class of graphical objects that the device is able to display. The space of graphical objects with this representation is called the *representation universe of the graphical display*. As an example, the representation universe of a regular personal computer (graphics card + monitor) is a matrix representation of an image. Thus, for each output graphics device, the operation of visualization on the device consists of an operator $R : O \rightarrow O'$ from the space of graphical objects to be displayed to the representation universe of the graphics display.

Besides the generic visualization operation described above, we have the operations of *graphical object description*, *graphical object processing*, and *graphical object analysis*. This completes the elements for obtaining a generic diagram for graphical objects, which contains, as particular cases, the diagrams in Figures 1.2 and 1.3.

1.4 DESCRIPTION, REPRESENTATION, AND RECONSTRUCTION

In this section, we discuss the three most important problems related to graphical object:

- How to describe a continuous graphical object?
- How to discretize a graphical object?

- How to obtain a continuous graphical object from one of its discretizations?

1.4.1 DESCRIPTION

Describing a graphical object consists in devising its (continuous) mathematical definition. Therefore, this is a problem posed in the mathematical universe. Depending on the nature of the graphical object, this topic is an important one in different areas of computer graphics. The description of the geometry and topology of objects is coped with geometric modeling, the description of motion is studied in the area of animation, and so on (See Figure 1.9).

Independent of the nature of the graphical object, there are essentially three mathematical techniques to describe a graphical object: parametric description, implicit description, and piecewise description. For more details about these methods, the reader should take a look at Velho *et al.* (2002).

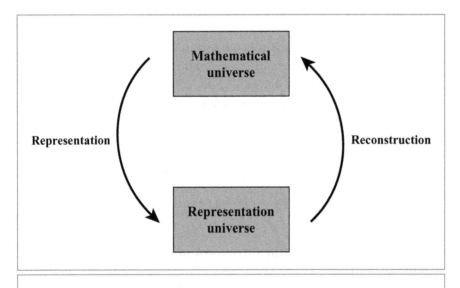

Figure 1.9: Representation and reconstruction.

1.4.2 REPRESENTATION

The representation of graphical objects constitutes a separate chapter in computer graphics. Because discretization is the essence of computational methods, representation techniques assume great importance in the area.

We have pointed out that the representation process implies a loss of information about the graphical object. In fact, representing a graphical object amounts to representing its geometric support and attribute function. Therefore, we may lose geometry and some attribute information.

From the mathematical point of view, we can formulate this problem in the following way. Given a space of graphical objects O_1 and a space of discrete graphical objects O_2, an operator $R : O_1 \rightarrow O_2$ is called a *representation operator*. This operator associates with each graphical object $x \in O_1$ its discrete representation $R(x) \in O_2$.

1.4.3 RECONSTRUCTION

An important problem consists in recovering the continuous graphical object in the mathematical universe from its representation. This is called the *reconstruction problem*.

Given an object $y \in O_2$, a *reconstruction* of y is a graphical object $x \in O_1$ such that $R(x) = y$. The reconstruction of y is denoted by $R^+(y)$. When R is an invertible operator, we have that $R^+(y) = R^{-1}(y)$. Note, however, that invertibility is not a necessary condition for reconstruction. In fact, it suffices that the operator R possesses a left inverse, $R^+(R(x)) = x$, to make reconstruction possible.

We should remark that the representation problem is a direct problem, while the reconstruction problem is an inverse problem.[2] A representation that allows more than one possibility of reconstruction is called

---

2   Direct and inverse problems are discussed in Chapter 1.

an *ambiguous representation.*[3] In this case, the reconstruction problem is ill posed (it has more than one solution).

Although the representation problem constitutes a direct problem, the computation of the representation operator could lead to an inverse problem. An example of such a situation can be seen in the point sampling representation of an implicit shape, $F^{-1}(0)$, where $F\colon \mathbb{R}^3 \to \mathbb{R}$. In this case, the representation consists in obtaining solutions to the equation $F(x, y, z) = 0$, which constitutes an inverse problem.

The representation takes a continuous graphical object O and associates with it a discrete representation O_d. The reconstruction process does the opposite, that is, from the discrete representation O_d, we obtain a reconstructed graphical object O_r. When $O_d = O_r$, we say that we have *exact reconstruction*, and the representation is also called *exact*. We should remark, however, that in general we do not have exact reconstruction, that is, in general, we obtain upon reconstruction only an approximation $O_r \simeq O$, and this approximation is enough for most of the applications.

The reader might be intrigued because we are talking about a continuous graphical object in the context of computer applications. How is it possible to have a continuous graphical object on the computer? Let us make this clear: we say that a graphical object $f\colon U \to \mathbb{R}^n$ is continuous when we are able access any point $x \in U$ and compute the value $f(x)$ of the attribute function at that point.

Example 5 [circle]. Consider the unit circle $S^1 = \{(x, y) \in \mathbb{R}^2; x^2 + y^2 = 1\}$ of the plane. A representation of S^1 can be obtained as follows. Take the parameterization $\varphi\colon [0, 1] \to S^1$, $\varphi(\theta) = (\cos 2\pi\theta, 2\pi \sin \theta)$, a partition $0 = \theta_1 < \theta_2 L \cdots < \theta_n$, and the representation is given by the sample vector $S_r^1 = (\varphi(\theta_1), \ldots, \varphi(\theta_n))$ (Figure 1.10(a) shows a representation for $n = 5$). A nonexact reconstruction of the circle

---

3   It would be more correct to adopt the term "ambiguous reconstruction," but the term "ambiguous representation" is already widely adopted in computer graphics.

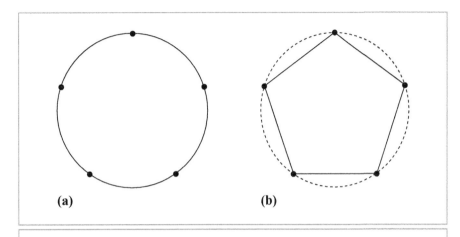

(a)   **(b)**

Figure 1.10: Uniform sampling of the circle with reconstruction.

from S_r^1 is given by the polygon with vertices $\varphi(\theta_1), \ldots, \varphi(\theta_n), \varphi(\theta_1)$ (Figure 1.10(b) illustrates this reconstruction for $n = 5$).

However, we know that the circle is completely defined by three distinct points; therefore, by taking a three-sample representation, we are able to obtain an exact representation of the circle.

We should remark that exact representations, although very useful in the solution to many problems, do not constitute the "saint graal." In fact, the polygonal representation of the circle, in spite of being non-exact, is very useful to visualize it (i.e., draw the circle on some output graphics device). In fact, drawing line segments is a very effective operation for any output graphical device; however, as the number of samples increase, the reconstructed polygon converges uniformly to the circle; thus, the reconstructed object provides a good approximation to the circle both visually and numerically.

Example 6 [terrain]. The uniform representation of a terrain (see Section 1.2.1) contains only height information of the terrain in a finite number of points. This implies a severe loss of information about the terrain. Is it possible to reconstruct the terrain data from their uniform representation? This is a very important question that can be restated as follows. Is it possible to devise an interpolation techique that allows us to interpolate the uniform samples of the height function in order to obtain the terrain height at any point (x, y) of the domain? Shannon theorem (see Gomes and Velho, 1997) provides a very general answer to this question.

Figure 1.11 shows two different reconstructions from the same terrain data. Note that the reconstructed terrain on the left has more geometric detail.

1.4.4 SEMANTICS AND RECONSTRUCTION

The reconstruction of a graphical object determines its semantics. Therefore, it is of great importance in the various processes of computer graphics. We can listen to a sound when it is reconstructed by a loudspeaker; we can see an image when it is reconstructed on the screen of a graphics display or printed on a paper. Therefore,

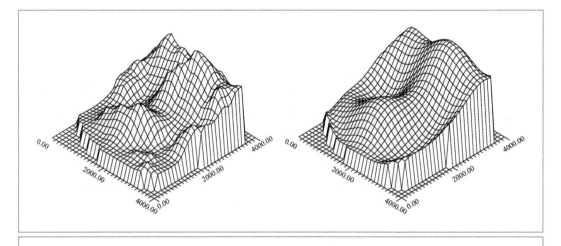

Figure 1.11: Different reconstructions from the same terrain data.

reconstruction techniques are present, at least, whenever we need to "display" information about the graphical object.

For the nonconvinced reader, we provide below four reasons to high-light the importance of reconstruction techniques:

1. When we need to obtain an alternative representation of an object, we can reconstruct it and then build a new representation from the reconstructed object.
2. The reconstruction is useful when we need to work in the continuous domain to minimize numerical errors.
3. In the visualization process, the output graphics device performs a reconstruction of the graphical object from its representation.
4. In general, the user describes a graphical object directly on some representation because a user interface allows only the input of a finite set of parameters. The object must be reconstructed from this "user representation."

In this way, reconstruction methods play a fundamental role in visualization. It is desirable to avoid ambiguous representations, which can make the reconstruction problem nonunique. A geometric example of an ambiguous representation is shown in Figure 1.12, where the object in image (a) is represented by a wireframe model. In images (b), (c), and (d), we can see three possible reconstructions of this object.

The reconstruction operation is totally dependent on the representation and, in general, it is difficult to be computed. Exact representations are rare. In the following chapters, we investigate how optimization techniques allow us to obtain good reconstructions of graphical objects.

1.5 COMMENTS AND REFERENCES

Until recently, many computer graphics books included a historical synopsis of the field's evolution. This started in early days when the field

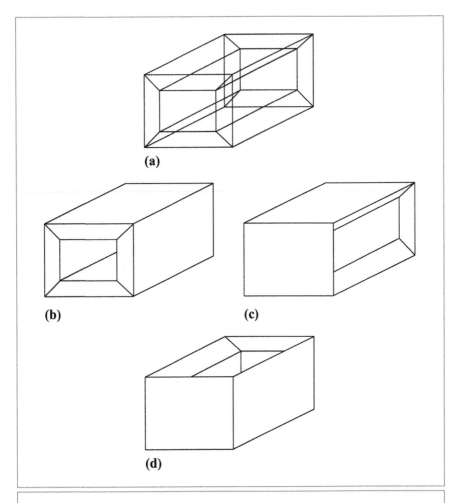

Figure 1.12: Ambiguous representation.

was new and relatively little known as a way to acquaint the public with the potential of computer graphics; later the tradition was maintained in a large measure because the explosive growth in the body of knowledge and applications demanded constant updating of the literature.

Today, although still a young discipline in comparison with other areas of science, computer graphics has developed to the point where the historical dimension plays a different role. A history of computer graphics must cover not only applications but also the evolution of mathematical and physical models, the algorithms, and even the hardware. Such a history, or even a bare chronology, would be far too long to be adequately dealt with in a single chapter. What is needed is a book entirely devoted to the history of computer graphics.

Nonetheless, we list below some work that has been done discussing Computer Graphics from a historical point of view.

The seminal work of Ivan Sutherland, Ph.D. thesis (Sutherland, 1963), marked a watershed between early rudimentary uses of the computer for graphics and the modern notion of interactive computer graphics. Sutherland's "Sketchpad" allowed the user to interactively manipulate plane geometric figures. The constraint-based modeling techniques used in the Sketchpad are among the first examples of optimization methods in graphics applications.

The first texts that refer to computer graphics as such were connected with computer-aided design (CAD) in high-technology industries, especially the automobile, aircraft, and shipbuilding industries (see, for example, Parslow, 1969, and Prince, 1971). An important issue in such systems is the design of fair surfaces that can be defined through variational optimization.

Although it can be appropriately applied to all areas of applied mathematics that involve computational methods, the four-universe paradigm first appeared explicitly in the literature in Requicha (1980) in the context of geometric modeling. More details on the use of this paradigm in various areas of computer graphics can be found in Gomes and Velho (1995).

As the subject of computer graphics matured, textbooks started to appear on specific subfields, such as ray tracing (Glassner, 1989) and lighting (Hall, 1989). The book (Fiume, 1989), devoted to raster graphics, is an effort to lay a solid conceptual foundation for this subject.

These textbooks cover to some extent graphic algorithms that use optimization methods.

BIBLIOGRAPHY

Fiume, E. L. *The Mathematical Structure of Raster Graphics*. San Diego, CA: Academic Press, 1989.

Glassner, A. (Ed). *An Introduction to Ray Tracing*. New York: Academic Press, 1989.

Gomes, J., L. Darsa, B. Costa, and L. Velho. Graphical objects. *The Visual Computer*, 12:269–282, 1996.

Gomes, J., and L. Velho. Abstraction paradigms for computer graphics. *The Visual Computer*, 11:227–239, 1995.

Gomes, J., and L. Velho. *Image Processing for Computer Graphics*. Secaucus, NJ: Springer-Verlarg, 1997.

Hall, R. A. *Illumination and Color in Computer Generated Imagery*. New York: Springer-Verlag, 1989.

Higham, N. J. *Accuracy and Stability of Numerical Algorithms*. Philadelphia, PA: SIAM Books, 1996.

Parslow, R. D., R. W. Prowse, and R. E. Green (Eds.) *Computer Graphics, Techniques and Applications*. London, UK: Plenum Press, 1969.

Prince, D. *Interactive Graphics for Computer Aided Design*. New York: Addison Wesley, 1971.

Requicha, A. A. G. Representations for rigid solids: Theory, methods, and systems. *ACM Computing Surveys*, 12(December):437–464, 1980.

Sutherland, I. *A Man-Machine Graphical Communication System*. Ph.D. Thesis, MIT Electrical Engeneering, 1963.

Velho, L., J. Gomes, and L. H. de Figueiredo. *Implicit Objects in Computer Graphics*. Secaucus, NJ: Springer-Verlarg, 2002.

Yamamoto, J. K. A review of numerical methods for the interpolation of geological data. *Anais da Academia Brasileira de Ciências*, 70(1):91–116, 1998.

2 OPTIMIZATION: AN OVERVIEW

2.1 WHAT IS OPTIMIZATION?

Intuitively, optimization refers to the class of problems that consists in choosing the *best* among a set of *alternatives*.

Even in this simple, imprecise statement, one can identify the two fundamental elements of an optimization problem: *best*, which conveys a choice of criterion used to choose the solution and is usually expressed by means of a function that should be minimized or maximized; and *alternatives*, which refers to the set of possible solutions that must be satisfied by any candidate solution. A simple example will help us clarify these remarks.

Example 7 [hotel-to-conference problem]. Find the best street to go from the hotel where you are staying to the convention center. The alternatives here consist of all the streets (or parts of streets) that when joined provide a path to go from your hotel to the convention center. It is clearly a finite set (provided that we avoid paths containing loops). We have different choices for the term *best*:

- the street that takes the smallest time;
- the shortest street;
- the street that has the best view of the landscape, etc.

From a mathematical viewpoint, an optimization problem may be posed as follows: *minimize the function* $f: S \rightarrow \mathbb{R}$, *that is, compute the pair* $(x, f(x))$ *such that* $f(x)$ *is the minimum value element of the set* $\{f(x); x \in S\}$. *We also use the two simplified forms below:*

$$\min_{x \in S} f(x) \quad \text{or} \quad \min\{f(x); x \in S\}.$$

The function $f: S \rightarrow \mathbb{R}$ is called *objective function*, and its domain S, the set of possible solutions, is called the *solution set*. We may restrict attention to minimization problems since any maximization problem can be converted into a minimization problem by just replacing f with $-f$. An *optimality condition* is a necessary or a sufficient condition for the existence of a solution to an optimization problem.

The nature of the solution set S is quite general as illustrated by the three examples below.

1. *The hotel-to-conference problem in Example 7.*
2. *Maximize the function* $f: S^2 \subset \mathbb{R}^3 \rightarrow \mathbb{R}$, *defined by* $f(x, y, z) = z$, *where* S^2 *is the unit sphere.*
3. *Find the shortest path that connects two points* p_1 *and* p_2 *of a surface* $M \subset \mathbb{R}^n$.

In the first problem, S is a finite set; in the second problem, $S = S^2$, a 2D surface of $\mathbb{R}^3$; in the third problem, S is the set of rectifiable curves of the surface joining the two points (in general, an infinite dimenional set).

2.1.1 CLASSIFICATION OF OPTIMIZATION PROBLEMS

The classification of optimization problems is very important because it will guide us to devise strategies and techniques to solve the problems from different classes. Optimization problems can be classified according to several criteria related to the properties of the objective function and also of the solution set S. Thus, possible classifications take into account

- the nature of the solution set S
- the description (definition) of the solution set S
- the properties of the objective function f.

The most important classification is the one based on the nature of the solution set S, which leads us to classify optimization problems into four classes: *continuous*, *discrete*, *combinatorial*, and *variational*.

A point x of a topological space S is an *accumulation point*, if and only if for any open ball B_x, with $x \in B_x$, there exists an element $y \in S$ such that $y \in B_x$. A point $x \in S$, which is not an accumulation point, is called an *isolated point* of S. Thus, x is isolated if there exists an open ball B_x, such that $x \in B_x$ and no other point of S belongs to B_x.

A topological space is *discrete* if it contains no accumulation points. It is *continuous* when all its points are accumulation points.

Example 8 [discrete and continuous sets]. Any finite subset of an Euclidean space is obviously discrete. The set $\mathbb{R}^n$ itself is continuous. The set $\mathbb{Z}^n = \{(i_1, \dots, i_n); i_j \in \mathbb{Z}\}$ is discrete. The set $\{0\} \cup \{1/n; n \in \mathbb{Z}\}$ is not discrete because 0 is an accumulation point (no matter how small we take a ball with center at 0, it will contain some element $1/n$ for some large n); this set is not continuous either because with the exception of 0, all the points are isolated. However, the set $\{1/n; n \in \mathbb{Z}\}$ (without the 0) is discrete.

2.2 CLASSIFICATION BASED ON THE NATURE OF SOLUTION

In this section, we discuss optimization methods according to the nature of the solution set.

2.2.1 CONTINUOUS OPTIMIZATION PROBLEMS

An optimization problem is called *continuous* when the solution set S is a continuous subset of $\mathbb{R}^n$.

The most common cases occur when S is a differentiable m-dimensional surface of $\mathbb{R}^n$ (with or without boundary), or $S \subseteq \mathbb{R}^n$ is a region of $\mathbb{R}^n$:

- Maximize the function $f(x,y) = 3x + 4y + 1$ on the set $S = \{(x,y) \in \mathbb{R}^2; x^2 + y^2 = 1\}$. The solution set S is the unit circle S^1, which is a curve of $\mathbb{R}^2$, a 1D surface (see Figure 2.1(a)).
- Maximize the function $f(x,y) = 3x + 4y + 1$ on the set $S = \{(x,y) \in \mathbb{R}^2; x^2 + y^2 \le 1\}$. The solution set is the disk of $\mathbb{R}^2$, a 2D surface with boundary (see Figure 2.1(b)).

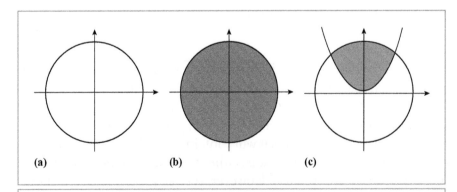

(a)   (b)   (c)

Figure 2.1: Three examples of a solution set.

- Maximize the function $f(x, y) = 3x + 4y + 1$ on the set S, defined by the inequalities $x^2 + y^2 \leq 1$ and $y - x^2 \geq 0$. The solution set is a region defined by the intersection of two surfaces with boundary (see Figure 2.1(c)).

2.2.2 DISCRETE OPTIMIZATION PROBLEMS

An optimization problem is called *discrete* when the solution set S is a discrete set (i.e., S has no accumulation points).

The most frequent case in the applications occurs for $S \subseteq \mathbb{Z}^n = \{(i_1, \ldots, i_n); i_n \in \mathbb{Z}\}$.

- Maximize the function $f(x, y) = x + y$ on the set $S = \{(x, y) \in \mathbb{Z}^2; 6x + 7y = 21, x \geq 0, y \geq 0\}$. The solution set S is a finite set, namely, the set of points with integer coordinates in a triangle whose sides are the axes and the line $6x + 7y = 21$ (see Figure 2.2(a)).

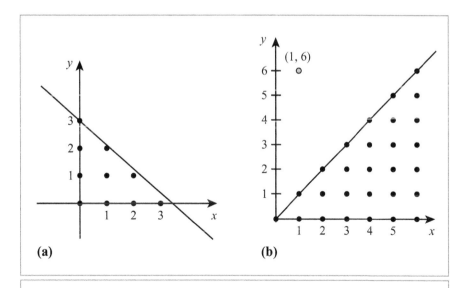

(a)   **(b)**

Figure 2.2: Discrete solution sets.

- Maximize the function $f(x, y) = (x - 1)^2 + (y - 6)^2$ on the set $S = \{(x, y); y \leq x, x \geq 0, y \geq 0\}$. The solution set S is infinite; the problem asks for finding among all points with integer coordinates in a cone, the one that is closest to the point $(1, 6)$ (see Figure 2.2(b)).

There is a close relationship between continuous and discrete optimization problems. In this respect, it is important to remark that some types of discrete problems $\min\{f(x); x \in S\}$ with solution set S may be solved more easily by embedding S into a continuous domain S', solving the new continuous optimization problem $\min\{f(x); x \in S'\}$, and obtaining the solution to the original problem from the solution in the continuous domain. The example below illustrates this approach.

Example 9. Consider the discrete optimization problem $\max\{f(x) = 3x - 2x^2; x \in \mathbb{Z}\}$. In order to solve this problem, we consider a similiar problem on $\mathbb{R}$, that is, $\max\{f(x) = 3x - 2x^2; x \in \mathbb{R}\}$.

The solution to the continuous problem is simple. In fact, since $f(x)$ is a quadratic function with negative second derivative, it has a unique maximum point given by $f'(x) = 0$, that is, $3 - 4x = 0$. Therefore, the solution is $m = 3/4$, and the maximum value is $9/8$. Now we obtain the solution to the discrete problem from this solution.

Note that the solution obtained for $S = \mathbb{R}$ is not a solution for the discrete case (because $m \notin \mathbb{Z}$). But it is possible to compute the solution for the discrete case from the solution to the continuous problem. For this, we just need to choose the integer number that is closest to m. It is possible to prove that this provides indeed the solution. In fact, since f is an increasing function in the interval $[-\infty, m]$ and decreasing function in the interval $[m, \infty]$ and its graph is symmetrical with respect to the line $x = m$, the furthest x is from m, the smaller is the value of $f(x)$. From this, we conclude that the integer n closest to m is indeed the point where f attains its maximum on the set $\mathbb{Z}$.

We should remark that in general the solution to a discrete problem cannot be obtained from a corresponding continuous problem in such a simple way as shown in the example above. As an example, consider the polynomial function $f: \mathbb{R} \to \mathbb{R}$, $f(x) = x^4 - 14x$. It has a minimum at $x \approx 1.52$. The closest integer value to this minimum point is 2, where f assumes the value -12. Nevertheless, the minimum of $f(x)$ when x is an integer number occurs for $x = 1$, where f attains the value -13.

Thus, although solving the continuous version of a discrete problem is part of the several existing strategies in computing a solution to a discrete problem, in general, finding the discrete solution requires more specific techniques that go beyond rounding off solutions.

2.2.3 COMBINATORIAL OPTIMIZATION PROBLEMS

An optimization problem is said to be *combinatorial* when its solution set S is finite. Usually, the elements of S are not explicitly determined. Instead, they are indirectly specified through combinatorial relations. This allows S to be specified much more compactly than by simply enumerating its elements.

In contrast to continuous optimization, whose study has its roots in classical calculus, the interest in solution methods for combinatorial optimization problems is relatively recent and significantly associated with computer technology. Before computers, combinatorial optimization problems were less interesting since they admit an obvious method of solution that consists in examining all possible solutions in order to find the best one. The relative efficiency of the possible methods of solution was not a very relevant issue: for real problems, involving sets with a large number of elements, any solution method, efficient or not, was inherently unfeasible for requiring a number of operations too large to be done by hand.

With computers, the search for efficient solution methods became imperative: the practical feasibility of solving a large-scale problem by computational methods depends on the availability of an efficient solution method.

Example 10 [the traveling salesman problem]. This need is well illustrated by the *traveling salesman problem*. Given n towns, one wishes to find the minimum length route that starts in a given town, goes through each one of the other towns, and ends in the starting town.

The combinatorial structure of the problem is quite simple. Each possible route corresponds to one of the $(n - 1)!$ circular permutations of the n towns. Thus, it suffices to enumerate these permutations, evaluate their lengths, and choose the optimal route.

This, however, becomes unpractical even for moderate values of n. For instance, for $n = 50$, there are approximately 10^{60} permutations to examine. Even if 1 billion of them were evaluated per second, examining all would require about 10^{51} seconds, or 10^{43} years! However, there are techniques that allow solving this problem, in practice, even for larger values of n.

2.2.4 VARIATIONAL OPTIMIZATION PROBLEMS

An optimization problem is called a *variational problem* when its solution set S is an infinite dimensional subset of a space of functions.

Among the most important examples, we could mention the *path* and *surface* problems. The problems consist in finding the best path (best surface) satisfying some conditions that define the solution set.

Typical examples of variational problems are the following.

Example 11 [geodesic problem]. Find the path of minimum length joining two points p_1 and p_2 of a given surface.

Example 12 [minimal surface problem]. Find the surface of minimum area for a given boundary curve.

Variational problems are studied in more detail in Chapter 4.

2.3 OTHER CLASSIFICATIONS

Other forms of classifying optimization problems are based on characteristics that can be exploited in order to devise strategies for the solution.

2.3.1 CLASSIFICATION BASED ON CONSTRAINTS

In many cases, the solution set S is specified by describing constraints that must be satisfyed by its elements. A very common way to define constraints consists in using equalities and inequalities. The classification based on constraints takes into account the *nature* of the constraints and the *properties* of the functions that describe them.

With respect to the nature of the constraint functions, we obtain the following classification:

- equality constraints: $h_i(x) = 0, i = 1, \ldots, m$;
- inequality constraints: $g_j(x) \leq 0$.

In the first case, the solution set S consists of the points that satisfy simultaneously the equations $h_i(x) = 0$. We should remark that in general the equation $h_i(x) = 0$ defines an m-dimensional surface of $\mathbb{R}^n$; therefore, the simultaneous set of equations $h_i(x) = 0$, $i = 1, \ldots, m$ define the intersection of m surfaces.

In the second case, each inequality $g_j(x) \leq 0$ in general represents a surface of $\mathbb{R}^n$ with boundary (the boundary is given by the equality $g_j(x) = 0$). Thus, the solution set is the intersection of a finite number of surfaces with boundary; in general, this intersection is a region of $\mathbb{R}^n$ (see Figure 2.1(c)). A particular case of great importance occurs when the functions g_j are linear. Then, the region is a solid polyhedron of $\mathbb{R}^n$.

As we will see in Chapter 5, equality and inequality constraints lead us to different algorithmic strategies when looking for a solution.

The algorithms are also affected by the properties of the functions that define the constraints. Among the particular and relevant special cases that can be exploited are the ones where the constraint functions are linear, quadratic, convex (or concave), or sparse.

Proposition below shows that the nature of the restriction functions influences the geometry of the solution set, and this geometry is widely exploited in the strategies to compute a solution to the problem.

Proposition 1. *If the restriction functions are convex, the solution set S is also convex.*

The result of this proposition can be exploited both to determine optimality conditions and to develop algorithms.

2.3.2 CLASSIFICATION BASED ON THE OBJECTIVE FUNCTION

The properties of the objective function are also fundamental in order to devise strategies for the solution to optimization problems. The special cases that lead to particular solution strategies are the ones where the objective function is linear, quadradic, convex (or concave), sparse, or separable.

Linear Programs

A very important situation occurs when both the objective and the constraint functions are linear and, moreover, the constraints are defined either by equalities or by inequalities, that is, the constraints are given by linear equations and linear inequalities. As we have remarked before, the solution set is a solid *polyhedron* of the Euclidean space. Optimization problems of this nature are called *linear programs*, and their study constitute a subarea of optimization. In fact, several practical problems can be posed as linear programs, and there exists a number of techniques that solve them, exploiting their characteristics: the linearity of the objective function and the piecewise linear structure of the solution set (polyhedra).

2.4 THE PROBLEM OF POSING PROBLEMS

Most of the problems in general and the problems of computer graphics in particular can be posed using the concept of operator between two spaces.[1] These problems can be classified into two categories: *direct* and *inverse* problems.

Direct problem.   Given are two spaces O_1 and O_2 of graphical objects, an operator $T : O_1 \rightarrow O_2$, and a graphical object $x \in O_1$. Problem: compute the object $y = T(x) \in O_2$. That is, we are given the operator T and a point x in its domain, and we must compute the image $y = T(x)$.

Because T is an operator, it is not multivalued; therefore, the direct problem always has a unique solution.

Inverse problems.   There are two types of inverse problems.

Inverse problem of first kind.   Given two spaces O_1 and O_2 of graphic objects, an operator $T : O_1 \rightarrow O_2$, and an element $y \in O_2$, determine an object $x \in O_1$ such that $T(x) = y$. That is, we are given an operator T and a point y belonging to the arrival set of the operator, and we must compute a point x in its domain whose image is y.

---

1   In some books the use of the term operator implies that it is linear. Here we use the term to mean a continuous function.

Inverse problem of second kind. Given a finite sequence of elements $x_1, x_2, \ldots, x_n \in O_1$ and a finite sequence $y_1, y_2, \ldots, y_n \in O_2$, determine an operator $T : O_1 \to O_2$ such that $T(x_i) = y_i$, $i = 1, 2, \ldots, n$. That is, we are given a finite sequence (x_i) of points in the domain and its image $(y_i) = (T(x_i))$, we must compute the operator T.

If the operator T is invertible with inverse T^{-1}, the inverse problem of first kind admits a unique solution $x = T^{-1}(y)$. However, invertibility is not necessary. In fact, a weaker condition occurs when T has a left inverse T^+ (i.e., $T^+ T = $ Identity), and the solution is given by $T^+(y)$. In this case, we cannot guarantee uniqueness of the solution because we do not have uniqueness of the left inverse.

The inverse problem of second type has two interesting interpretations:

1. We are given an operator $\overline{T} : \{x_1, \ldots, x_n\} \to \{y_1, \ldots, y_n\}$, and we must compute an extension T of $\overline{T}$ to the entire space O_1. Clearly, in general, this solution, if it exists, is not unique.
2. The solution operator T defines an interpolation from the samples $(x_i, T(x_i))$, $i = 1, \ldots, n$.

From the above interpretations, it is easy to conclude that, in general, the solution to any type of inverse problems is not unique.

2.4.1 WELL-POSED PROBLEMS

Hadamard[2] has established the concept of a *well-posed* problem as a problem in which the following conditions are satisfied:

1. There is a solution.
2. The solution is unique.
3. The solution depends continuously on the initial conditions.

---

2   Jacques Hadamard (1865–1963), French mathematician.

When at least one of the above conditions is not satisfied, we say that the problem is *ill posed*.

We have seen previously that, in general, inverse problems of the first and second kinds are ill posed in the sense of Hadamard because of the nonuniqueness of the solution.

The concept of an ill-posed problem needs to be further investigated very carefully. The continuity condition is important because it guarantees that small variations in the initial conditions will cause only small perturbations of the solution. Recall that, in practical problems, small variations in the initial conditions are quite common (and sometimes, inevitable) due to measurement imprecision or numerical errors. The nonuniqueness of solutions, however, should not be necessarily an obstacle. An example will make this point clear.

Example 13. Consider the operator $F : \mathbb{R}^2 \to \mathbb{R}$ defined by $F(x, y) = x^2 + y^2 - 1$ and the inverse problem of the first kind $F(x, y) = 0$. That is, we must solve the quadratic equation $x^2 + y^2 - 1 = 0$. It is clear that it admits an infinite number of solutions and therefore it is ill posed in the sense of Hadamard. Geometrically, these solutions constitute the unit circle S^1 of the plane. The nonuniqueness of the solutions is very important in the representation of the circle. In fact, in order to obtain a sampled representation of the circle, we need to determine a finite subset of these solutions. Figure 2.3 shows the use of seven solutions of the equation that allows us to obtain an approximate reconstruction of the circle by an heptagon.

The use of the term ill posed to designate the class of problems that do not satisfy one of the three Hadamard conditions may induce the reader to conclude that ill-posed problems cannot be solved, but this is far from the truth. Interpreting ill-posed problems as a category of intractable problems, the reader would be discouraged to study computer graphics, an area in which there is a large number of ill-posed problems, as we will see in this chapter.

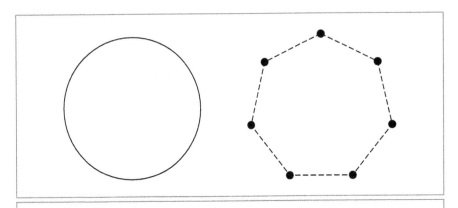

Figure 2.3: Polygonal representation of a circle.

In fact, ill-posed problems constitute a fertile ground for the use of optimization methods. When a problem presents an infinite number of solutions and we wish to have unicity, optimization methods make it possible to reformulate the problem, so that a unique solution can be obtained. The general principle is to define an objective function such that an "optimal solution" can be chosen among the set of possible solutions. In this way, when we do not have a unique solution to a problem, we can have the best solution (which is unique).

It is interesting to note that, sometimes, well-posed problems may not present a solution due to the presence of noise or numerical errors. The example below shows one such case.

Example 14. Consider the solution of the linear system

$$x + 3y = a$$
$$2x + 6y = 2a,$$

where $a \in \mathbb{R}$. In other words, we want to solve the inverse problem $TX = A$, where T is the linear transformation given by the matrix

$$\begin{pmatrix} 1 & 3 \\ 2 & 6 \end{pmatrix},$$

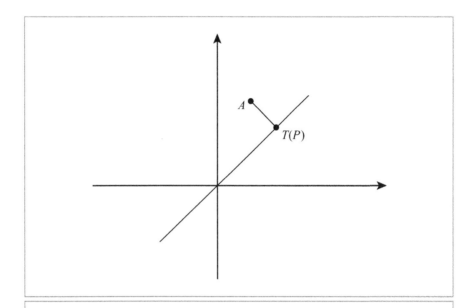

Figure 2.4: Well-posed formulation of a problem.

$X = (x, y)$ and $A = (a, 2a)$. It is easy to see that the image of $\mathbb{R}^2$ by T is the line r given by the equation $2x - y = 0$. Because the point A belongs to the line r, the problem has a solution (in fact, an infinite number of solutions).

However, a small perturbation in the value of a can move the point A out of the line r, and the problem fails to have a solution.

To avoid this situation, a more appropriate way to pose the problem uses a formulation based on optimization: *determine the element $P \in \mathbb{R}^2$ such that the distance from $T(P)$ to the point A is minimal.* This form of the problem always has a solution. Moreover, this solution will be close to the point A. Geometrically, the solution is the point P, such that its image $T(P)$ is the closest point of the line r to the point A (see Figure 2.4).

2.4.2 PROBLEM REDUCTION

Another important concept related to problem characterization is the notion of *problem reduction*. It constitutes a powerful conceptual device that makes possible to analyze and compare problems in the same class. Problem reduction also has practical relevance because it enables finding the right algorithm to resolve a problem.

Imagine you have a problem that you do not know how to solve. Very often, using a reduction technique, you can turn this problem into a prototypical one that can be solved. This not only gives you the methodology for implementing the solution but usually comes with guarantees associated with the knowledge of the problem nature.

A variation in the above strategy is when you can decompose a complex problem that you do not know how to solve into the aggregate solution of simpler subproblems that you can solve. This leads to some of the basic algorithm design methodologies in computer science, such as the *divide and conquer* technique.

Moreover, if your problem can be reduced to any instance of a known hard problem, then you can be sure that your problem is at least as hard. This issue will be discussed in more detail in the next section.

2.5 HOW TO SOLVE IT?

It is the purpose of this book to use optimization techniques to solve computer graphics problems.

Later, we present several examples of computer graphics problems that can be posed, in a natural way, as optimization problems. Nevertheless, we should remind that posing a problem as an optimization problem does not mean that it is automatically solved, at least in the sense of obtaining an exact solution.

In fact, optimization problems are, in general, very difficult to solve. Some of the difficulties are discussed below.

2.5.1 GLOBAL VERSUS LOCAL SOLUTIONS

In continuous optimization problems, most of the optimality conditions are related to local optimality problems. The computation of global optimality solutions is, in general, very expensive from the computational point of view. Thus, in practice, we have to settle for the best local optimal solution found by the algorithm.

2.5.2 NP[3]-COMPLETE PROBLEMS

There are combinatorial problems—for example, the so-called *NP-complete* problems—for which there are no efficient methods (i.e., running in polynomial time on the size of the instance to be solved) that provide exact solutions in all cases. The traveling salesman problem, mentioned before, is in that class. These problems are related in such a way that if an algorithm with polynomially bounded running time is found for one of the problems, then polynomial time algorithms will be available for all. This makes it unlikely that such an algorithm will ever be found. When solving such problems, many times, we have to use heuristic solution processes that provide good (but not optimal) solutions.

2.6 COMMENTS AND REFERENCES

The bibliography about optimization is vast, and it is not our intention to go over a complete review of the area. In this section, we suggest some textbooks that cover, with more details, the topics we have discussed in this chapter.

---

3 Complexity class of "nondeterministic polynomial time" problems.

Luenberger (1984) provides a broad overview of continuous optimization, covering both linear and nonlinear problems. For discrete and combinatorial optimization, refer to Papadimitriou and Steiglitz (1982). Weinstock (1974) gives an excellent introduction to variational optimization, with emphasis on applications.

There are many books in the literature dedicated to linear programming and its applications to combinatorial problems (and in particular for net flows). Chvatal (1983) is a good example. Fang and Puthenpura (1993) is a more recent book, which emphasizes the most recent algorithms of interior points for linear programming. Gill *et al.* (1981) mostly deal with nonlinear optimization, emphasizing the practical aspect of choosing the most appropriate algorithm for some given optimization problem.

BIBLIOGRAPHY

Chvatal, V. *Linear Programming*. W. H. Freeman, 1983.

Fang, S.-C., and S. Puthenpura. *Linear Optimization and Extensions: Theory and Algorithms*. Upper Saddle River, NJ: Prentice Hall, 1993.

Gill, P. E., W. Murray, and M. H. Wright. *Practical Optimization*. San Diego, CA: Academic Press, 1981.

Luenberger, D. G. *Linear and Nonlinear Programming*. Norwell, MA: Addison-Wesley, 1984.

Papadimitriou, C. H., and K. Steiglitz. *Combinatorial Optimization: Algorithms and Complexity*. Upper Saddle River, NJ: Prentice-Hall, 1982.

Weinstock, R. *Calculus of Variations with Applications to Physics and Engineering*. Dover: Weinstock Press, 1974.

3 OPTIMIZATION AND COMPUTER GRAPHICS

In this chapter, we show the importance of optimization methods in computer graphics. We pose several problems in the area, whose solutions use optimization techniques. The problems we pose here will be solved in the following chapters with the appropriate optimization techniques: continuous, discrete, combinatorial, or variational.

3.1 A UNIFIED VIEW OF COMPUTER GRAPHICS PROBLEMS

Computer graphics problems are, in general, posed by considering two spaces O_1 and O_2 of graphical objects and relations between them.[1] In several practical situations, the spaces O_1 and O_2 are topological spaces, and the relation is an operator (i.e., a continuous function)

---

1 A relation is a subset of the Cartesian product $O_1 \times O_2$.

from O_1 to O_2. An even more well-behaved condition occurs when O_1 and O_2 are vector spaces and the operator is linear.

We consider the intermediate case when we have two topological spaces of graphical objects O_1 and O_2 and an operator $R : O_1 \rightarrow O_2$ from O_1 to O_2.

From this, we have the following generic problems:

- For a given graphical object $x \in O_1$, compute $T(x)$ (direct problem).
- Given $y \in O_2$, compute the graphical object $T^{-1}(y)$ (inverse problem).
- Given graphical objects x, y, with $x \in O_1$ and $y \in O_2$, compute the operator T such that $T^{-1}(y)$ (inverse problem).

In this chapter, we study several specific cases of the above problems.

3.1.1 THE CLASSIFICATION PROBLEM

Classification problems, also called *discrimination problems*, allow the recognition of graphical objects based on the identification of the class to which they belong.

In this section, we give a generic description of the classification problem and demonstrate how it leads to optimization methods in a natural manner.

A relation $\equiv$ defined on a set O of graphical objects is an *equivalence relation* if it satisfies the following conditions:

1. $O_\alpha \equiv O_\alpha$ (reflexivity)
2. $O_\alpha \equiv O_\beta \Rightarrow O_\beta \equiv O_\alpha$ (symmetry)
3. $O_\alpha \equiv O_\beta$ and $O_\beta \equiv O_\theta \Rightarrow O_\alpha \equiv O_\theta$ (transitivity).

The *equivalence class* $[x]$ of an element $x \in O$ is the set $[x] = \{y \in O; y \equiv x\}$. From the properties of the equivalence relation, it follows that two equivalence classes coincide or are disjoint. Moreover, the union of all equivalence classes is the set O. Thus, the equivalence classes constitute a *partition* of the space O of graphical objects.

Thus, whenever we have an equivalence relation in a set O of graphical objects, we obtain a classification technique, which consists in obtaining the corresponding partition of the set O.

How to obtain a classification on a set O of graphical objects? A powerful and a widely used method consists in using a *discriminating function*, which is a real function $F : O \rightarrow U \subset \mathbb{R}$. The equivalence relation $\equiv$ in O is defined by

$$x \equiv y \quad \text{if and only if} \quad F(x) = F(y).$$

It is easy to prove that $\equiv$ is indeed an equivalence relation. For each element $u \in U$, the inverse image $F^{-1}(u) = \{x \in O; F(x) = u\}$ defines an equivalence class of the relation.

When we have a classification of a set O of graphical objects, a natural problem consists in choosing an *optimal* representative, $\bar{u}$ in each equivalence class $F^{-1}(u)$. For this, we define a distance function $d : O \times O \rightarrow \mathbb{R}$ on the space O.[2] Then, we choose the best representative element $\bar{u}$ of the class $[u]$ as the element that minimizes the function

$$\sum_{o \in [u]} d(o, \bar{u}).$$

An interesting case occurs when the set O of graphical objects is finite, $O = \{o_1, \ldots, o_m\}$. A classification amounts to dividing O in a partition with n sets. $[u_1], \ldots, [u_n]$, $n < m$ (in general, n much smaller than m). This problem can be naturally posed as an optimization problem: *What is the optimal partition of the space O?*

---

2   A distance function differs from a metric function because it does not necessarily satisfy the triangle inequality.

In some classification problems, it is possible to define a probability distribution p on the set O of graphical objects. In this case, a more suitable measure for an optimal partition would be

$$E = \sum_{i=1}^{m} \sum_{o \in [\bar{u}]} p(o)d(o, \bar{u}).$$

The function E depends on the partition in classes and the choice of representative for each class. The minimization of the function E over all possible partitions of the set O is a problem of combinatorial optimization. This problem is known in the literature as *cluster analysis*.

Example 15 [database queries]. In the area of multimedia databases (e.g., image databases), the classification methods play a fundamental role in the problem of indexation and query of objects in the database.

3.2 GEOMETRIC MODELING

Large part of the problems in modeling is related to the description, representation, and reconstruction of geometric models. As we have already mentioned, there are essentially three methods to describe a geometric shape:

1. Implicit description
2. Parametric description
3. Piecewise description (implicit or parametric).

The description is usually given in a functional form, either deterministic or probabilistic.

A model is called *procedural* when its describing function is an algorithm over some virtual machine.

In general, the geometry of a model is specified by the user through a finite number of parameters; based on this specification, the model is reconstructed.

Some common cases are the following:

- Reconstruction of a curve or a surface from a set of points. In this case, the reconstruction uses some interpolation method for scattered data.
- Reconstruction of surfaces based on a finite set of curves.

It is clear that the two problems above lack uniqueness of solution. Additional conditions are required for unicity. These conditions can be obtained by algebraic methods or by using optimization. In the first case, we may impose that a curve or a surface belong to some specific class, for example, defined by a polynomial of degree 3. In the second case, we may look for a curve or a surface that minimizes some given objective function.

3.2.1 MODEL RECONSTRUCTION

This is a very important problem in geometric modeling. In general, a model is specified by a finite number of parameters that represent its geometry, topology, and attributes.

An important task consists in reconstructing the model from these finite number of parameters.

1. The user specifies a finite number of parameters using some input device.
2. We obtain samples from a terrain data.
3. Samples from a computerized tomography (CT) or magnetic resonance imaging (MRI).

As specific cases of this model, we have the problems of *curve modeling* and *surface modeling*. The methods to create models using optimization are known in the literature as *variational modeling*.

3.3 VISUALIZATION

The visualization process consists in the generation of images from geometric models. Formally, we define a *rendering operator* that associates each point in the ambient space with a corresponding point in the image with its respective color.

The rendering operator, $R : O_1 \rightarrow O_2$, maps graphical objects from the space of scene O_1 to graphical objects on the space of the visualization device O_2. Usually, O_1 is the space of 3D objects, and O_2 is the space of 2D regions of the screen.

It is possible to formalize the complete rendering pipeline in terms of operators on graphical objects. John Hart has used this method to analyze the hardware rendering pipeline on OpenGL accelerated graphics cards (Hart *et al.*, 2002).

However, the solution to this direct problem is not simple because the rendering operator depends on many factors: light sources, camera transformation, geometry, and material properties of the objects of the scene and ultimately of the propagation of radiant energy through the scene reaching the image plane. Depending on the type of illumination model adopted, the rendering computation can be a direct or an inverse problem.

There are many optimization problems that arise from the inverse computation of the rendering operator. For example, we may wish to determine where to position light sources in a scene such that the shadows they cast into the objects have a certain effect.

Also, effective representation of the rendering operator is the fundamental factor for interactive and real-time visualization. In this respect, optimization techniques for finding the most efficient and compact representations, such as factorization of the reflectance functions, and precomputation of light transport play a key role.

3.4 COMPUTER VISION

In this area, we are looking for physical, geometrical, or topological information about the objects of a scene that is depicted in an image. In a certain way, the image is a representation of the scene, and the vision problem consists in the reconstruction of the 3D scene from its 2D representation. In the visualization process that generates an image, there is a great loss of information implied by the projection that changes the dimensionality of the representation. The image, therefore, is an extremely ambiguous representation.

As described above, computer vision is clearly an inverse problem.

Even the reconstruction of the scene by our sophisticated visual system may present ambiguities. Two classic examples of such a semantic ambiguity are shown in Figure 3.1. In (a), we can see an image

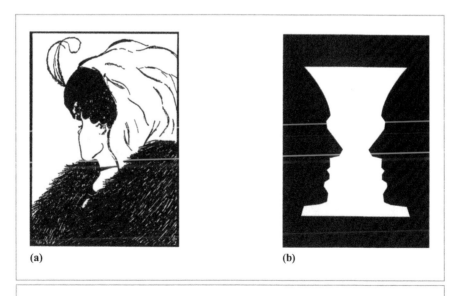

(a)   (b)

Figure 3.1: Ambiguous reconstructions.

that can be interpreted as the face of either an elderly woman or a young lady (the nose of the old woman is the chin of the lady). In (b), we can interpret the figure either as the silhouette of two faces over a white background or as a white glass over a black background. Note that the ambiguity of the second image is related to the problem of distinguishing between foreground and background in the reconstruction.

The complexity of reconstruction in computer vision originates different versions of the reconstruction problem. Two important cases of the problem are as follows.

- **Shape from shading:** Obtain the geometry of the objects based on information about the radiance of the pixels in the image.
- **Shape from texture:** Obtain information about the geometry of the scene based on a segmentation and an analysis of the textures in the image.

3.5 THE VIRTUAL CAMERA

The virtual camera is the main link between the areas of visualization and vision of 3D scenes.

It is the element that maps 3D objects in the scene into 2D information in the image, and it is used in a dual manner in visualization and vision processes.

The simplest model of a virtual camera is the *pinhole camera*, which has 7 degrees of freedom representing the parameters

- position (3 degrees of freedom)
- orientation (3 degrees of freedom)
- focal distance (1 degree of freedom).

These parameters determine a projective transformation T that associates with each point $p \in \mathbb{R}^3$ a point in the image plane (for more details, see Gomes and Velho, 1998). In other words, given a point $p \in \mathbb{R}^3$, the virtual camera transformation generates the point $T(p) = p'$ in the image.

We have two inverse problems related in a natural way to the virtual camera:

1. Given a point p' in the image, determine the point p of the scene, assuming that the camera parameters are known.
2. Given points p and p', determine the camera transformation T such that $T(p) = p'$.

In the first problem, we could have many points in the scene, which correspond to the point p' in the image (i.e., 3D points that are projected in the same 2D image point). In the second case, it is clear that the knowledge of just one 3D point and its 2D projection is not sufficient to determine the seven camera parameters.

One important problem is the application of point (2) to the case of images produced by real cameras. In other words, given an image acquired by some input device (video camera or photographic camera), determine the camera parameters (position, orientation, focal length, etc.). This problem is known as *camera calibration*.

Camera calibration has a large number of applications:

- In virtual reality applications, it is often necessary to combine virtual images with real ones. In this case, we need to synchronize the parameters of the virtual camera with those of the real camera in order to obtain a correct alignment of the elements in the image composition.
- TV broadcast of soccer often uses 3D reconstruction of a given play to help spectators better understand what has happened. In order to do this, it is necessary to know the camera parameters. More information about this type of application can be obtained from http://www.visgraf.impa.br/juizvirtual/.

3.5.1 CAMERA SPECIFICATION

An important problem in graphics is the camera specification. In the direct specification, the user provides the seven camera parameters that are required to define the transformation T. Then, we have the direct problem of computing image points $p' = T(p)$.

A related inverse problem arises in the study of user interfaces for camera specification. In general, as we have seen, the user specifies the transformation T through the seven camera parameters, leading to an easy-to-solve direct problem. However, this interface is not appropriate when the user wants to obtain specific views of the scene (e.g., keeping the focus of attention on a certain object). A suitable interface should allow the user to frame an object directly on the image. Such a technique is called *inverse camera specification*.

In the inverse specification, the 7 degrees of freedom of the camera is specified *indirectly* by the user. The user wishes to obtain a certain framing of the scene, and the specification must produce the desired result.

The idea of the inverse specification is to allow the definition of camera parameters inspired on the "cameraman paradigm": the user (in the role of a film director) describes the desired view directly on the image, and the system adjusts the camera parameters to obtain that view.

The inverse camera specification facilitates the view description by the user at the cost of having to solve an ill-posed mathematical problem. Let us see how it works using a concrete example.

Example 16 [framing a point in the image]. Consider the situation illustrated in Figure 3.2. Point P in space is fixed, and the observer is at point O; this point is projected in point A on the screen. The user requests a new view, subject to the constraint that point A is at the center of the image. Therefore, we need to obtain camera parameters such that point P will now be projected in point B that is located at the center of the screen.

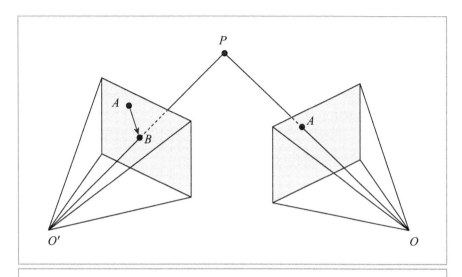

Figure 3.2: Inverse specification to frame a point in the image (Gomes and Velho, 1998).

Figure 3.2 shows a new position O' and a new orientation of the camera that solves the problem. Observe that this camera setting is not a unique solution to the problem.

From the mathematical point of view, we have a transformation $T : \mathbb{R}^7 \to \mathbb{R}^2$, where $y = T(x)$ is the projection of the parametrization space of the camera on the Euclidean plane. The solution to our problem is given by the inverse $T^{-1}(B)$. However, the transformation T is not linear, and, in general, it does not have an inverse (why?). Therefore, the solution has to be computed using optimization techniques. That is, we need to look for the "best" solution in some sense.

3.6 IMAGE PROCESSING

Image processing studies different operations with images. These operations are characterized by operators in spaces of images.

Among the many problems that need to be solved in image processing, we can highlight the classification problem. Two important cases of classification problems are image segmentation and image quantization. In the first case, classification is posed on the geometric support of the image, and in the second case, classification is done in the range of the image function (i.e., the color space).

3.6.1 WARPING AND MORPHING

An important operation consists in the warping of the domain U of an image $f : U \subset \mathbb{R}^2 \to \mathbb{R}$. More specifically, given an image $f : U \to \mathbb{R}$, a warping of U is a diffeomorphism $g : V \subset \mathbb{R}^2 \to U$. This diffeomorphism defines a new image given by $h : V \to \mathbb{R}, h = f \circ g^{-1}$. An example of such an operation is illustrated in Figure 3.3.

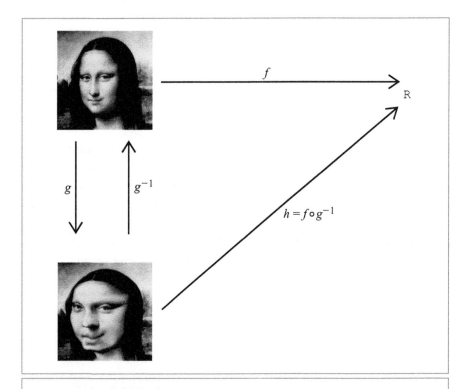

Figure 3.3: Image deformation.

Image warping has many applications, such as image morphing, registration, and correction.

Image morphing.   It deforms an image into another in a continuous way (see Figure 3.4).

Registration of medical images.   The comparison of different medical images is very common in diagnosis. It is also common to combine images of the same part of the body obtained from different sources. In all these cases, the images need to be perfectly aligned.

Correcting image distortions.   Many times, the process of capturing or generating an image introduces distortions. In such cases, it is necessary to correct the distortions with a warping. We use a standard pattern to measure distortions, and based on those measures, we correct other images produced by the process. Figure 3.5 shows an example of a standard pattern and its deformation due to some image capture process.

One of the main problems in the area of warping and morphing is the specification of the warping operator. There are many techniques for this purpose. A simple and a very popular method is the punctual specification (Figure 3.6). In order to deform the bird into the dog, we select some points in the outline curve of the bird and their corresponding points in the outline of the dog. Through the knowledge of the deformation acting on this set of points, it is possible to reconstruct the desired deformation over the whole space.

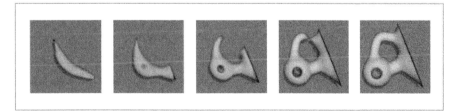

Figure 3.4: Image metamorphosis.

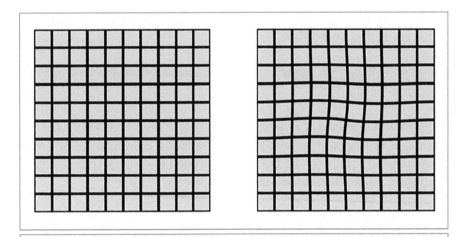

Figure 3.5: Correcting distortions.

Figure 3.6: Punctual specification.

With punctual specification, the problem of finding a deformation reduces to the problem of determining a warping $g : \mathbb{R}^2 \to \mathbb{R}^2$ such that

$$g(p_i) = q_i, \quad i = 1, \ldots, n,$$

where p_i and q_i are points of $\mathbb{R}^2$. It is clear that this problem does not admit a unique solution, in general.

Suppose that $g(x, y) = (g_1(x, y), g_2(x, y))$, where $g_1, g_2 : U \to \mathbb{R}$ are the components of g. In this case, the equations $g(p_i) = q_i$ can be written in the form

$$g_1(p_x^i, p_y^i) = q_x^i;$$
$$g_2(p_x^i, p_y^i) = q_y^i,$$

with $i = 1, \ldots, n$. That is, the problem amounts to determining two surfaces

$$(x, y, g_1(x, y)), \quad (x, y, g_2(x, y)),$$

based on the knowledge of a finite number of points on each surface. Note that we encounter the same problem in the area of modeling.

The way that the deformation problem was posed using point specification is too rigid. If we restrict the class of warpings that are allowed, the problem may not have a solution. In addition, the above formulation makes the problem very sensitive to noise. A more convenient way to pose the problem consists in forcing $g(p_i)$ close to p_i. In this new formulation, the solution is less sensitive to noise, and it is possible to restrict the class of deformations and still obtain a solution. The new problem is, therefore, an optimization problem.

Particularly interesting cases of the formulation above consist in considering space deformations induced by rigid motions or linear or projective transformations. A survey discussing several approaches to solve the problem using rigid motions can be seen in Goodrich *et al.* (1999). A comprehensive treatment of warping and morphing of graphical objects can be found in Gomes *et al.* (1998).

3.7 IMAGE ANALYSIS

Image analysis deals with the inference of models from images. These techniques are used in vision and pattern recognition. Depending on the level of the model, it can fall into the category of low-level, midlevel, or high-level vision. In this section, we give two examples: one is a low-level problem (edge detection) and the other a midlevel problem (character recognition).

3.7.1 EDGE DETECTION

A classical problem in the area of image analysis is edge detection. Intuitively, edges are the boundaries of objects in the image. This is shown in Figure 3.7, where the white pixels in image (b) indicate the edges of image (a).

The first researcher to call attention to the importance of edge detection in image analysis was D. Marr.[3] He posed an important problem in this area, which became known as the *Marr's Conjecture: an image is completely characterized by its edges*. In order to pose this problem more precisely, it is necessary to define the concept of edge, but we do not define here. We resort, instead, to a high-level discussion of the problem.

Edge detection methods can be divided into two categories:

- Frequency-based methods
- Geometric methods.

Frequency-based methods characterize edges as regions of high frequency of the image (i.e., regions where the image function exhibits large variations). The theory of operators over images is used to determine those regions. The reconstruction of an image from its edges is, therefore, related to the invertibility of these operators. The wavelet transform plays a significant role.

Geometric methods try to describe edges using planar curves over the image support. This approach uses largely optimization methods. A classic example is given by the *snakes*.

---

3 David Marr (1945–1980), English scientist and MIT researcher, was one of the pioneers in the field of computer vision.

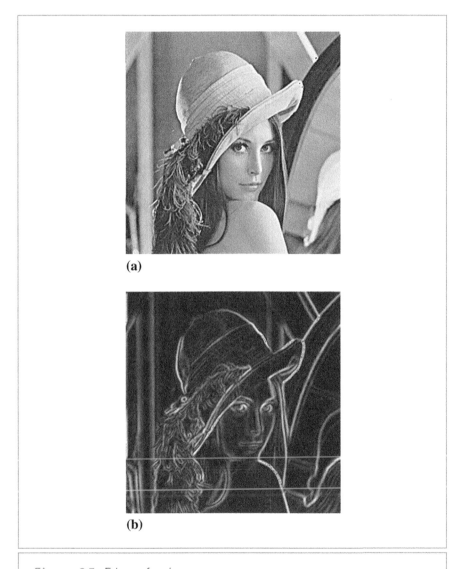

(a)

(b)

Figure 3.7: Edges of an image.

Snakes.   These curves reduce the edge detection problem to a variational problem of curves in the domain of the image. We use the image function to define an external energy in the space of curves. This external energy is combined with an internal energy of stretching and bending in order to create an energy functional. The functional is such that minimizer curves will be aligned with the edges of regions in the image. The computation of snakes is usually done using a relaxation process starting with an initial curve near the region. Figure 3.8 illustrates edge detection using snakes: (a) the initial curve and (b) the final curve fully aligned with the edges of a region.

3.7.2 CHARACTER RECOGNITION

A classification problem of practical importance is the automatic character recognition. Given a binary image of a scanned text, identify

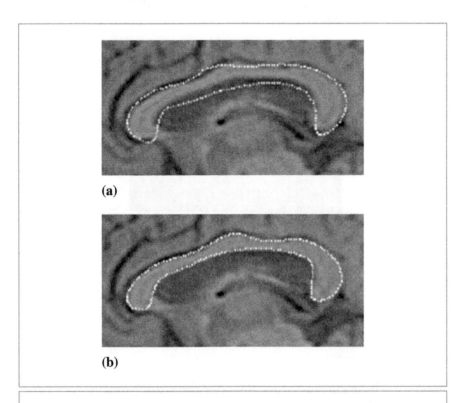

(a)

(b)

Figure 3.8: Edge detection using snakes: (a) initial curve; (b) final edge curve (Brigger and Unser, 1998).

all the printed glyphs of the text. This problem has applications in optical character recognition systems, which transform a scanned image in a text that can be edited in a word processor.

Note that each character represents a region of the plane, which can be interpreted as a planar graphical object. We need to identify the associated letter of the alphabet for each of these regions. If we define the depiction of a character as another planar graphical object, then the problem is to find a discriminating function to measure the similarity of this type of objects (Atallah, 1984). After that, we proceed as in the general classification problem described above.

3.8 ANIMATION AND VIDEO

The area of animation is concerned with the problem of time-varying graphical objects. There are three important problems in animation:

- Motion synthesis
- Motion analysis
- Motion processing.

In general, motion synthesis is a direct problem, and motion analysis leads to various types of inverse problems.

Motion synthesis is closely related to the area of motion control and uses optimization methods extensively (e.g., optimal control).

Motion visualization, or animation, is done through a sequence of images, which constitute essentially a temporal sampling of the motion of the objects in a scene. Such an image sequence is also known as *digital video*. Note that this opens up a perspective for new problems: analysis and processing of video.

A particularly interesting problem consists in recovering the motion of 3D objects in the scene based on the analysis of their 2D motion in the video, which is known as *motion capture*. This is another example of a highly ill-posed problem in the sense of Hadamard, where optimization methods are largely used to find adequate solutions.

3.9 COMMENTS AND REFERENCES

The importance of mathematical optimization methods for computer graphics and vision has been apparent since the early developments in these fields. It has gained increasing importance as the fields evolved and the problems to be solved became more complex. Perhaps, the awareness of the need for optimization tools reached its peak with the integration of graphics and vision applications in areas such as *image-based modeling and rendering* (Szeliski *et al.*, 1998).

BIBLIOGRAPHY

Atallah, M. Checking similarity of planar figures. *International Journal of Computing Information Science*, 13:279–290, 1984.

Brigger, P., R. Engel, and M. Unser. B-spline snakes and a JAVA interface: An intuitive tool for general contour delineation. In *Proceedings of IEEE International Conference on Image Processing, Chicago, IL, USA, October 4–7*, 1998.

Gomes, J., L. Darsa, B. Costa, and L. Velho. *Warping and Morphing of Graphical Objects*. San Francisco, CA: Morgan Kaufmann, 1998.

Gomes, J., and L. Velho. *Computação Gráfica, Volume 1*. Série de Computação e Matemática, Rio de Janeiro, RJ, Brasil: IMPA, 1998.

Goodrich, M. T., J. S. B. Mitchell, and M. W. Orletsky. Approximate geometric pattern matching under rigid motions. *PAMI*, 21(4):371–379, 1999.

Hart, J., M. Olano, W. Heidrich, and M. McCool. *Real Time Shading*. AK Peters, Ltd., 2002.

Szeliski, R., M. Levoy, P. Hanrahan, L. McMillan, and E. Chen (Eds). *Workshop on Image-Based Modeling and Rendering*. Stanford University, ACM Press, New York, NY: 1998.

4 VARIATIONAL OPTIMIZATION

4.1 VARIATIONAL PROBLEMS

A large number of the optimization problems that occur in computer graphics are variational problems. In those problems, the set of possible solutions is the space of functions defined on a continuous domain.

The most classical variational problem is the following:

Problem 0. Find a function $\mathbf{y} : [x_1, x_2] \to R^n$, satisfying the conditions $\mathbf{y}(x_1) = \mathbf{y}_1$ and $\mathbf{y}(x_2) = \mathbf{y}_2$, that minimizes the integral $\int_{x_1}^{x_2} f(x, \mathbf{y}', \mathbf{y})dx$, where f is a known real function.

The optimality conditions for the problem above are expressed by a system of partial differential equations, given below.

Theorem 1 [Euler–Lagrange conditions]. *Let* $\mathbf{y} : [x_1, x_2] \to R^n$ *be a twice-differentiable function that solves the problem above. Then,* $\mathbf{y}$ *must satisfy the following system of partial differential equations:*

$$\frac{\partial f}{\partial y_i} = \frac{\partial}{\partial x}\left(\frac{\partial f}{\partial y_i'} \right), i = 1, \ldots, n.$$

In most cases, one cannot solve these equations analytically. For this reason, the solution to these problems are usually based on numerical techniques, which require some kind of discretization. There is no general rule to obtain this discretization, and different techniques of numerical methods for partial differential equations are found in the literature. We use the example below to discuss possible solution strategies.

Example 17. Let us consider the problem of finding a minimum length curve joining points $P_1 = (x_1, y_1, g(x_1, y_1))$ and $P_2 = (x_2, y_2, g(x_2, y_2))$ on a surface in $\mathbb{R}^3$ having equation $z = g(x, y)$ (see Figure 4.1).

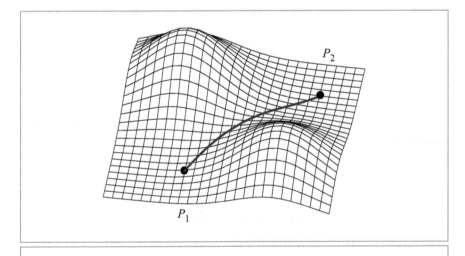

Figure 4.1: Minimum length path on a surface.

Let $u : [0, 1] \to R^2$ be a parametrization of the curve. Its length can be written as $\int_0^1 ||u'(t)||dt$. Therefore, the problem to be solved is that of minimizing this integral subject to the condition that $u(0) = P_1$ and $u(1) = P_2$. Hence, this is a specific case of the problem above.

With little modification, this mathematical formulation applies to problems that are apparently quite different in nature. For instance, consider the problem of detecting a segment of the boundary (between two given points) of the region corresponding to the brain, in the image in Figure 4.2. An image can be seen as a surface $z = g(x, y)$, where z is the gray level at position (x, y).

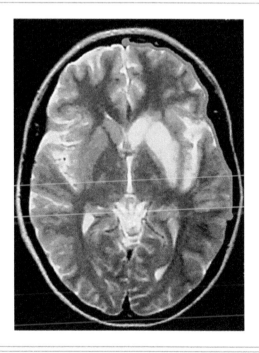

Figure 4.2: Detecting a boundary segment.

The boundary segment can again be found by minimizing an appropriate functional of the form

$$\int_0^1 ||u'(t)||w(u(t))dt \qquad (4.1)$$

over all curves $u(t)$ joining the two given points. The w factor should be small at the boundary points in order to make them more attractive when minimizing the functional. Observe that the case of the minimum length curve is a specific instance of the this formulation, for $w(u) = 1$.

For boundary detection, we observe that the magnitude of the gradient tends to be large at points on the boundary. Thus, for the case at hand, choosing $w(u) = \frac{1}{1+|\nabla f(u)|}$ gives good results.

Once the problem has been formulated as a variational problem, we face the task of finding a solution. There are three possible approaches.

4.1.1 SOLVE THE VARIATIONAL PROBLEM DIRECTLY

This requires solving directly the partial differential equation that expresses the optimality conditions for the problem. Even for this simple case, the Euler–Lagrange conditions result in a system of differential equations that cannot be solved analytically, except for special cases (for instance, when the surface is a plane).

Since analytical solutions are not available, the equations must be solved numerically, which demands the introduction of some discretization scheme.

4.1.2 REDUCE TO A CONTINUOUS OPTIMIZATION PROBLEM

This strategy consists in projecting the infinite dimensional solution space into a space of finite dimension. In general, the solution obtained is only an approximation to the solution to the original problem. As an example, we could consider the problem of minimizing the length of parametric cubic curves, which are defined by the the equation

$$\alpha(t) = (c_{10} + c_{11}t + c_{12}t^2 + c_{13}t^3, c_{20} + c_{21}t + c_{22}t^2 + c_{23}t^3).$$

This reduces our search to an 8D space of the parameters c_{ij} on which we can use methods and techniques from continuous optimization. We should remark, however, that even in this continuous problem, we must resort to some discretization in order to compute, approximately, the integral that provides the length of the curve.

4.1.3 REDUCE TO A DISCRETE OPTIMIZATION PROBLEM

Here, the discretization occurs in the beginning of the process. The domain of the objective function f is discretized using a regular grid, and we consider the graph G whose vertices are the grid intersections, whose edges correspond to pairs of vertices, and the length of each edge is the length, on the surface geometry, of the curve generated by the line segment that connects the two vertices (see Figure 4.3).

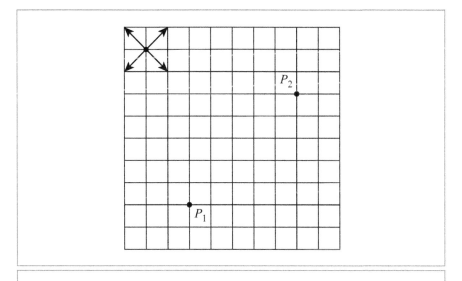

Figure 4.3: Discretizing a variational problem.

The curve of minimal length is found by solving a problem of minimum path in the graph G (see Chapter 6). The quality of the approximation obtained by this method certainly depends on the resolution of the grid (number of vertices). We should remark, however, that the more we increase the grid resolution, the greater the computational cost of the solution.

4.2 APPLICATIONS IN COMPUTER GRAPHICS

In this section, we study the problem, from computer graphics, that can be modeled as a variational problem: variational modeling of curves.

This problem originates from the area of *variational modeling*, which consists in using variational techniques in the area of geometric modeling.

4.2.1 VARIATIONAL MODELING OF CURVES

The basic problem of describing curves is related to the creation of geometric objects in the context of specific applications. A generic method consists in determining a curve that passes through a finite set of points $p_1, \ldots, p_n$ and at the same time satisfy some application-dependent criteria. In general, besides obtaining a curve with a given shape, many other conditions that measure the "quality" of the curve can be imposed, such as continuity, tangent line at control points, and continuous curvature.

Each condition imposes some constraint on the curve to be constructed. These constraints can be of analytical, geometric, or topological nature. A convenient way to obtain curves that satisfy some set of conditions is to pose the problem in the context of optimization: define an energy functional such that the curves that are minimizers of this functional automatically satisfy the desired criteria.

D. Bernoulli[1] was the first mathematician to propose a functional to measure the energy associated with the tension of a curve. This energy is called *tension energy* and is proportional to the square of the curvature k of the curve:

$$E_{\text{tension}}(\alpha) = \int_{\alpha} k^2 ds, \qquad (4.2)$$

where s is the arc length of the curve α. Note that the tension energy of a straight line is zero because its curvature is identically null. However, if the tension energy of a curve is high, then it "bends" significantly. Curves that minimize the tension energy are called *nonlinear splines*.

The geometric constraints that are imposed can be of a punctual or a directional nature. An example of punctual constraint is to force the curve to pass through a finite set of points. An example of directional constraint is to force the tangent vector of the curve to have a certain direction, previously specified.

The geometric constraints are, in general, modeled by an energy functional associated with external forces that act on the curve. The tension energy discussed earlier is an internal energy to the curve. In this way, we have a unification of the optimization problem looking for the minima of an energy functional E that has an internal component and an external component:

$$E_{\text{total}}(\alpha) = E_{\text{int}}(\alpha) + E_{\text{ext}}(\alpha).$$

A simplification adopted very often consists in decomposing the internal energy into a *bending* and a *stretch* component. In this way, the internal energy is given by the linear combination

$$E_{\text{int}}(\alpha) = \mu E_{\text{bending}} + (1 - \mu)E_{\text{stretch}},$$

---

1   Daniel Bernoulli (1700–1782), Swiss mathematician known for his work in hydrodynamics and mechanics.

$\mu \in [0, 1]$. The bending energy is given by Equation (4.2), and the stretch energy is given by

$$E_{\text{stretch}}(\alpha) = \int_{\alpha} ||\alpha'(t)|| \, dt.$$

Therefore, the internal energy controls the energy accumulated by the curve when it is bent or stretched. A minimization of this functional results in a curve that is as short and straight as possible. If the curve is constrained to pass through two points, the solution to the minimization problem is certainly a line segment. The external energy results in nontrivial solutions to the problem.

Intuitively, the external energy deforms the curve in many ways, bending and stretching it. In general, the user specifies various components of external energy, which are linearly combined in order to obtain the desired deformation. This energy is defined in the form of attractors or repulsors, which can be of two kinds: *positional* and *directional*. We give below two examples to clarify these concepts.

Example 18 [punctual attractors]. An example of a positional attractor is the punctual attractor, whose energy functional is given by

$$E_{\text{punctual}}(\alpha) = d(\alpha, p)^2 = \min_t ||\alpha(t) - p||^2.$$

Figure 4.4(a) shows an example of a curve that passes through points p_0 and p_1 while being deformed by attraction to point p.

Example 19 [directional attractor]. An example of a directional attractor is based on specifying a line $r(t) = p + tv$. The curve starts at point p_0 and ends at point p_1. The external energy makes the curve to approach point p and the tangent to the curve at points near p will line up with vector v (see Figure 4.4(b)). This condition can be obtained by an energy functional given by the equation

$$E_{\text{dir}}(\alpha) = \min_t ||\alpha'(t) \wedge v||^2.$$

According to the classification in Chapter 2, the problem of variational modeling of curves consists in finding a curve satisfying certain

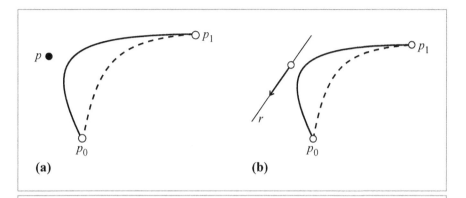

Figure 4.4: Attractors: punctual (a) and directional (b).

constraints (e.g., pass through a finite number of points $p_1, p_2, \ldots, p_n$) that minimize a certain energy functional (the objective function). This functional is defined in such a way that the curves that minimize it have some of the desired characteristics, such as being smooth and staying close to some prespecified points.

In order to fix the ideas, we consider the case of finding a curve α that passes through $p_1, p_2, \ldots, p_n$ and minimizes the energy functional

$$E = E_{\text{int}} + E_{\text{ext}}, \tag{4.3}$$

where

$$E_{\text{int}}(\alpha) = \mu E_{\text{tension}} + (1 - \mu)E_{\text{stretch}} = \mu \int_{\alpha} k^2 ds + (1 - \mu) \int_{\alpha} |\alpha'(t)|\, dt$$

and

$$E_{\text{ext}} = c \sum_{j=1}^{m} \min_{t} d(\alpha(t), q_j)^2.$$

In the above equations, μ and c are constants, and $\{q_0, q_1, \ldots, q_m\}$ is a set of attractors.

This is a variational problem. That is, the solution set has infinite dimension, and its elements consist of all the smooth curves that pass through the given points. An approach to solve the problem consists in using the Euler–Lagrange equation to formulate it as a partial derivative problem. Another approach, which we adopt here, is that of solving an approximation to the problem by projecting it onto an appropriate finite dimensional space.

Wesselink (1996) has used the space of *uniform cubic B-spline* curves for this purpose. These curves are described by the equations

$$\alpha(t) = \sum_{i=0}^{n-1} P_i N_i^3(t), \quad t \in [2, n-1], \tag{4.4}$$

where $P_0, P_1, \ldots, P_n$ are called *control points* and $N_1^3, N_2^3, \ldots, N_{n-1}^3$ constitute a *basis* of cubic uniform B-splines (see Bartels *et al.*, 1987) defined by

$$N_i^0(u) = \begin{cases} 1, & \text{if } i-1 \leq t < i \\ 0, & \text{otherwise} \end{cases}$$

$$N_i^d(u) = \frac{t-i+1}{d} N_i^{d-1}(t) + \frac{i+d-t}{d-1} N_{i+1}^{d-1}(t).$$

It is possible to prove that for every $t \in [2, n-1]$, we have $\sum_{i=1}^{n-1} N_i^3(t) = 1$. Moreover, $N_i^3(t) \geq 0$ $(i = 1, \ldots, n-1)$. Therefore, each point $\alpha(t)$ of the curve is a convex combination of control points. The basis functions, whose graphs are depicted in Figure 4.5., determine the weight of each control point in $\alpha(t)$ (Figure 4.6).

There are several reasons to justify the choice of the space of cubic B-spline curves:

- Cubic B-splines are frequently used in interactive modeling of curves (mainly because of the properties described below). This

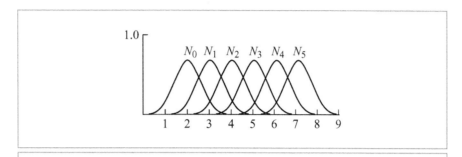

Figure 4.5: Basis of B-spline curves.

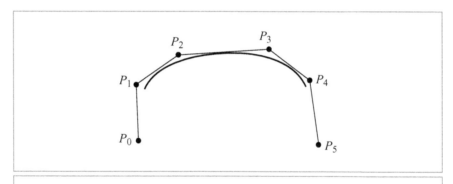

Figure 4.6: Control points and B-spline curves.

enables us to add variational modeling techniques to the existing modeling systems.

- Cubic B-splines are piecewise polynomial curves, with the smallest degree in such a way as to provide an adequate level of continuity for most of the applications (of first order in the entire curve and of second order except at a finite number of points).

- Cubic B-splines have local control. That is, each basis function has compact support, and as a consequence, each control point influences only part of the curve that is very useful. In an interactive modeling, it implies that changes in control points affect only part of the curve. However, it results from the compactness of the support that the interpolation constraints are expressed by a sparse system of equations.

The problem of finding a curve given by Eq. 4.4, which minimizes the functional given in Eq. 4.3, is still difficult from the computational point of view because the terms of external energy involve integrals for which there are no analytical expression available. Moreover, the constraints associated with the requirement that the curve contain a set of points, as formulated above, do not specify the values of the parameter t corresponding to those points. If the variables corresponding to these parameter values are introduced in the problem, such constraints become nonlinear. An analogous difficulty occurs when computing the attraction energy.

Thus, it is necessary to introduce additional simplifications:

- The energies of stretching and bending can be approximated by quadratic functions of the control points, using

$$E_{\text{bending}} \approx \int_{\alpha} |\alpha(t)''|^2 dt$$

and

$$E_{\text{stretching}} \approx \int_{\alpha} |\alpha(t)'|^2 dt.$$

Both integrals above are of the form $\sum_{i,j} a_{ij} P_i P_j$, where the constants a_{ij} are integrals expressed in terms of the basis functions.
- In the interpolation constraints and in the attraction functionals, we specify the corresponding parametric values. This forces that the interpolation constraints become linear as a function of the

control points and the attraction functional becomes a quadratic function of the same points.

Thus, the problem is reduced to the one of minimizing a quadratic function of the control points subject to linear constraints, and it can be solved using the first-order optimality condition.

The modeling process described in Wesselink (1996) is the following:

1. An initial curve is modeled in the conventional way using the manipulation of the control points.
2. On this curve, the user specifies the interpolation points and outside of the curve the user specifies the attractor points.
3. The control points are then recomputed in such a way to minimize the total energy according to the the interpolation constraints.
4. The user may insert new control points or adjust the curve manually and return to step 2 above.

Figure 4.7 shows, on the right, the result of specifying two attractors to the curve on the left, which was required to pass through the two extreme points and the central point and maintain the vertical direction at the other two marked points.

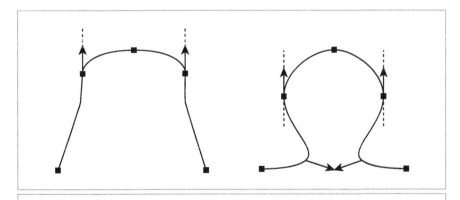

Figure 4.7: Interactive variational modeling.

4.3 COMMENTS AND REFERENCES

The bibliography on variational methods is quite extensive and can be mathematically challenging. We refer the reader to Weinstock (1974) for an introductory text on variational optimization, with emphasis on applications.

BIBLIOGRAPHY

Bartels, R., J. C. Beatty, and B. Barsky. *An Introduction to Splines for Use in Computer Graphics and Geometric Modeling.* San Fransisco, CA: Morgan and Kaufman, 1987.

Weinstock, R. *Calculus of Variations with Applications to Physics and Engineering.* Weinstock Press, Dover: 1974.

Wesselink, J. W. *Variational Modeling of Curves and Surfaces.* Ph.D. Thesis, Technische Universiteit Eindhoven, 1996.

5 CONTINUOUS OPTIMIZATION

In this chapter, we study the continuous optimization methods and techniques. As we know from Chapter 2, continuous optimization problems are posed in the form

$$\min_{x \in S} f(x),$$

where f is a real function and S is a subset of $\mathbb{R}^n$, described using a, possibly empty, set of equations and inequalities. Thus, the general problem that we study is

$$\begin{aligned} \min \quad & f(x) \\ \text{subject to} \quad & h_i(x) = 0, \quad i = 1, \ldots, m \\ & g_j(x) \geq 0, \quad j = 1, \ldots, p. \end{aligned}$$

We pay special attention to the case where $m = p = 0$. In this case, the solution set is the entire space $\mathbb{R}^n$, and the optimization problem is called *unconstrained*.

Consider a continuous optimization problem $\min_{x \in \mathbb{R}^n} f(x)$ and a point $x_0 \in \mathbb{R}^n$. The methods and techniques studied in this chapter provide answers to the following questions:

- Is x_0 a solution to the optimization problem?
- If not, how can we compute a better alternative than the one provided by x_0?

As we will see, the answers are extensions, to $\mathbb{R}^n$, of the well-known classical theory of maxima and minima for real functions of real variables. The main tool used in the proofs is *Taylor's theorem*, which provides a polynomial approximation to a function $f : \mathbb{R}^n \to \mathbb{R}$ in a neighborhood of a point a using the value of f and its derivatives at the point a.

In particular, we use the second-order approximation given by the infinitesimal version of Taylor's theorem (see Rudin, 1976).

Theorem 2. *Consider a function $f : \mathbb{R}^n \to \mathbb{R}$ and a point $x_0, h \in \mathbb{R}^n$. Suppose that f is twice differentiable at x_0. Define $r : \mathbb{R}^n \to \mathbb{R}$ by*

$$f(x_0 + h) = f(x_0) + \nabla f(x_0) \cdot h + \frac{1}{2} h^\top \nabla^2 f(x_0) \cdot h + r(h).$$

Then, $\lim_{h \to 0} r(h)/|h| = 0.$

In the above theorem, ∇f and $\nabla^2 f$ denote, respectively, the *gradient* vector and the *Hessian* matrix of the function f, defined by

$$\nabla f(x) = \left(\frac{\partial f}{\partial x_1}(x) \cdots \frac{\partial f}{\partial x_n}(x) \right)$$

and

$$\nabla^2 f(x) = \begin{pmatrix} \frac{\partial^2 f}{\partial x_1 \partial x_1}(x) & \cdots & \frac{\partial^2 f}{\partial x_1 \partial x_n}(x) \\ \vdots & \ddots & \vdots \\ \frac{\partial^2 f}{\partial x_n \partial x_1}(x) & \cdots & \frac{\partial^2 f}{\partial x_n \partial x_n}(x) \end{pmatrix}.$$

5.1 OPTIMALITY CONDITIONS

The theorems that follow establish necessary and sufficient conditions in order that $x_0 \in \mathbb{R}^n$ be a local minimum or maximum point of a function f.

We start with the theorem that establishes for functions of n variables that the derivative must vanish at the local extrema.

Theorem 3 [first-order necessary conditions]. *Let $f : \mathbb{R}^n \to \mathbb{R}$ be a function of class C^1. If x_0 is a local minimum (or maximum) point of f, then $\nabla f(x_0) = 0$.*

Proof. Suppose that $\nabla f(x_0) \neq 0$. Take an arbitrary vector d such that $\nabla f(x_0) \cdot d < 0$ (e.g., we may choose $d = -\nabla f(x_0)$). We have from the first-order Taylor's formula (or simply from the definition of derivative)

$$f(x_0 + \lambda d) = f(x_0) + \lambda \nabla f(x) \cdot d + r(\lambda)$$
$$= f(x_0) + \lambda(\nabla f(x_0) \cdot d + \frac{r(\lambda)}{\lambda}),$$

where $r(\lambda)$ is such that $\lim_{\lambda \to 0} \frac{r(\lambda)}{\lambda} = 0$. For $\lambda > 0$ and sufficiently small, $\nabla f(x_0) \cdot d + \frac{r(\lambda)}{\lambda} < 0$, which implies that $f(x_0 + \lambda d) < f(x_0)$, and this shows that x_0 is not a local minimum. By taking $-d$ instead of d, we show, in the same way, that x_0 is not a local maximum either. $\square$

The above proof is based on an important idea, which is used by several algorithms: if $\nabla f(x_0) \neq 0$, then it is possible, starting from x_0, to find direction vectors d such that the value of f decreases along these directions (at least locally). These directions, which are characterized by satisfying the inequality $\nabla f(x_0)d < 0$, are called *descent directions* (see Figure 5.1). Note that the condition $\nabla f(x_0) = 0$ expresses the fact that there does not exist either descent or ascent directions at x_0.

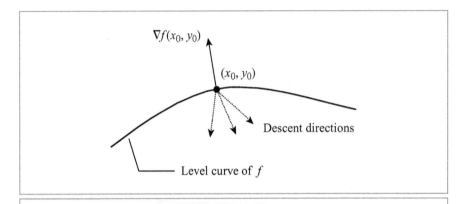

Figure 5.1: Descent directions.

For this reason, the points where this condition is satisfied are called *stationary points* of f.

We should remark that the condition stated in Theorem 3 is only necessary. The points that satisfy these conditions, i.e., the points where the gradient vector is null are called *critical points* of the function f. As in the 1D case, it is possible to use the second derivative of f at x_0 in order to obtain more information about the critical point x_0.

Theorem 4 [second-order necessary conditions]. *Let* $f : \mathbb{R}^n \to \mathbb{R}$ *be a function of class* C^2. *If* x_0 *is a local minimum of* f, *then* $\nabla f(x_0) = 0$ *and the Hessian matrix* $H = \nabla^2 f(x_0)$ *is nonnegative (i.e., the Hessian matrix satifies* $d^{\mathsf{T}} H d \geq 0$ *for every* $d \in \mathbb{R}^n$).

Proof. Let x_0 be a local minimum of f. From the first-order conditions, we have $\nabla f(x_0) = 0$. Suppose then there exists a direction vector d such that $d^{\mathsf{T}} H d < 0$. From Theorem 2, we have

$$f(x_0 + \lambda d) = f(x_0) + \lambda \nabla f(x_0) \cdot d + \frac{1}{2}\lambda^2 d^{\mathsf{T}} H d + r(h)$$

$$= f(x_0) + \lambda^2 \left(\frac{1}{2} d^{\mathsf{T}} H d + \frac{r(\lambda)}{\lambda^2} \right),$$

where $\lim_{\lambda \to 0} \frac{r(\lambda)}{\lambda^2} = 0$. Since $d^{\mathsf{T}} H d < 0$, it follows that $d^{\mathsf{T}} H d + \frac{r(\lambda)}{\lambda^2} < 0$. Therefore, we have $f(x_0 + \lambda d) < f(x_0)$ for $|\lambda|$ sufficiently small, and this contradicts the fact that x_0 is a local minimum. Therefore, we must have $d^{\mathsf{T}} H d \geq 0$ for every d. □

The condition $d^{\mathsf{T}} H d \geq 0$ for every d is analogous to the nonnegativeness of the second derivative at a local minimum of a real function of one real variable. We must impose extra conditions on H in order to guarantee that x_0 is indeed a local minimum of f. In the 1D case, it is sufficient to impose that the second derivative be positive at x_0; in a similar way, in the case of n variables, we must demand that the Hessian matrix be positive.

Theorem 5 [second-order necessary and sufficient conditions].
Let $f : \mathbb{R}^n \to \mathbb{R}$ be a function of class C^2. Suppose that $\nabla f(x_0) = 0$ and that the Hessian matrix $H = \nabla^2 f(x_0)$ is positive (i.e., $d^{\mathsf{T}} H d > 0$, for every $d \neq 0$). Then, x_0 is a strict local minimum of f.

Proof. Let x_0 be a point of $\mathbb{R}^n$ satisfying the conditions of the theorem. Since $d^{\mathsf{T}} H d > 0$ for every $d \neq 0$, there exists $a > 0$ such that $d^{\mathsf{T}} H d > a|d|^2$ for every $d \neq 0$ (it is enough to consider a as being the maximum value of $d^{\mathsf{T}} H d$ in the compact set $\{d \in \mathbb{R}^n \,|\, |d| = 1\}$). Then, from the second-order Taylor's formula,

$$f(x + d) = f(x_0) + f(x_0) \cdot d + \frac{1}{2} d^{\mathsf{T}} H d + r(d)$$

$$> f(x_0) + \frac{1}{2} a|d|^2 + r(d)$$

$$= f(x_0) + |d|^2 \left(\frac{a}{2} + \frac{r(d)}{|d|^2} \right),$$

where r is such that $\lim_{d \to 0} \frac{r(d)}{|d|^2} = 0$. Hence, for $|d|$ sufficiently small, we have $a + \frac{r(d)}{|d|^2} > 0$ and as a consequece, we obtain $f(x + d) > f(x_0)$. Therefore, f has a strict local minimum at x_0. □

Summarizing this discussion, if x_0 is a critical point of f (i.e., such that $\nabla f(x_0) = 0$) and H is the Hessian matrix of f at x_0, then we have the following cases:

a) If $d^\mathsf{T} H d > 0$ for every $d \neq 0$, then x_0 is a relative minimum point of f.

b) If $d^\mathsf{T} H d < 0$ for every $d \neq 0$, then x_0 is a relative maximum point of f.

c) If there exist d_1 and d_2 such that $d_1^\mathsf{T} H d_1 > 0$ and $d_2^\mathsf{T} H d_2 < 0$, then x_0 is neither a relative minimum nor a relative maximum of f. In fact, f has a minimum at x_0 along the direction d_1 and a maximum at x_0 along the direction d_2. In this case, we say that x_0 is a *saddle point* of f.

d) If $d^\mathsf{T} H d \geq 0$ for every d (with $d_1^\mathsf{T} H d_1 = 0$ for some direction $d_1 \neq 0$ and $d_2^\mathsf{T} H d_2 > 0$ for some direction d_2), then f *may* have a local minimum at x_0 (and certainly does not have a local maximum there). Similarly, if $d^\mathsf{T} H d \leq 0$ for every d (with $d_1^\mathsf{T} H d_1 = 0$ for some direction $d_1 \neq 0$ and $d_2^\mathsf{T} H d_2 < 0$ for some direction d_2), then f *may* have a local maximum at x_0 (and certainly does not have a local minimum there).

It is possible to illustrate all the cases described above using quadratic functions of two variables. That is, functions of the form $f(x) = c + b^\mathsf{T} x + \frac{1}{2} x^\mathsf{T} H x$, where b and c are 2×1 vectors and H is a symmetric 2×2 matrix. It is easy to verify that the gradient of f is $\nabla f(x) = (b + Hx)^\mathsf{T}$ and that its Hessian matrix is $\nabla^2 f(x) = H$. Therefore, the critical points of f are the solutions to the equation $Hx = -b$. If H is not singular, then there exists a unique critical point $x_0 = -H^{-1}b$. In this case, the occurrence of a maximum or a minimum point at this critical point depends on the positivity or negativity of the matrix H, that is, on the sign of its eigenvalues. If H is singular, f may or may not have critical points. The examples below illustrate these situations.

a) H has two positive eigenvalues (e.g., if $f(x, y) = 2x + x^2 + 2y^2$). In this case, f has a unique local minimum at the point $x_0 = -H^{-1}b = (1, 0)$. The level curves of f are illustrated in

Figure 5.2. In fact, f has a global minimum at x_0 since one can easily verify that

$$f(x) = \frac{1}{2}(x - x_0)^\top H(x - x_0) + f(x_0).$$

(Analogously, if H has two negative eigenvalues, then f has a unique local maximum, which will also be a global maximum).
b) H has a positive and a negative eigenvalue (e.g., if $f(x, y) = 2x + x^2 - 2y^2$). In this case, the unique critical point of f (which is $x_0 = -H^{-1}b = (1, 0)$) is a saddle point and f does not have

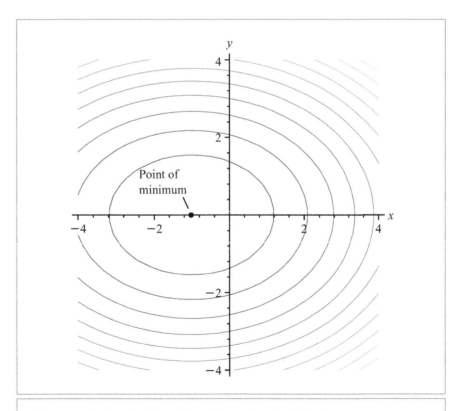

Point of minimum

Figure 5.2: Quadratic function with a unique local minimum.

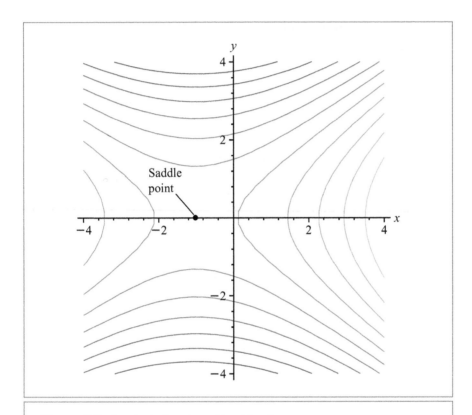

Figure 5.3: Quadratic function without local extrema.

relative extrema. Figure 5.3 shows the level curves of f for this case.

c) H has a positive and a null eigenvalue. If b does not belong to the 1D space generated by the columns of H, then f does not have critical points and therefore does not have relative extrema (Figure 5.4). This is the case for the function $f(x) = 2x + 2y^2$. However, if b belongs to the space generated by the columns of H, then the set of solutions to the equation $Hx_0 = -b$ is a straight line. Each point x_0 of this line is a critical point of f. Moreover, since $f(x) = (x - x_0)^\mathsf{T} H(x - x_0) + f(x_0)$, each one of these critical points is a local (and global) minimum point of f. (This illustrates the fact that the sufficient conditions of second

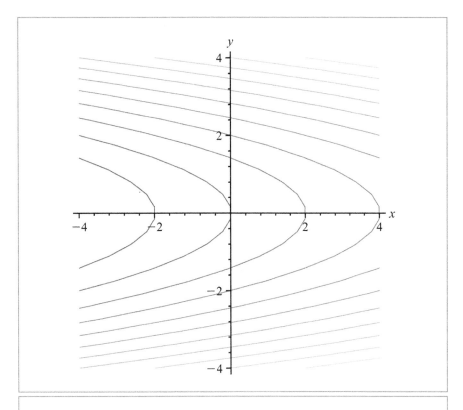

Figure 5.4: Quadratic function without local extrema.

order are not necessary conditions for optimality: H is not positive at the critical points, but the latter are local minima). As an example, for $f(x) = 2x + x^2$, the minima are the points of the line $x = 1$. Figure 5.5 illustrates this last situation.

5.1.1 CONVEXITY AND GLOBAL MINIMA

The theorems of the previous section provide necessary and sufficient conditions to verify the occurrence of local minima or maxima of a function. In practice, however, we are interested in *global* maxima and minima. General techniques to obtain these extrema are studied in Chapter 7. However, there exists a class of functions, called *convex*, for

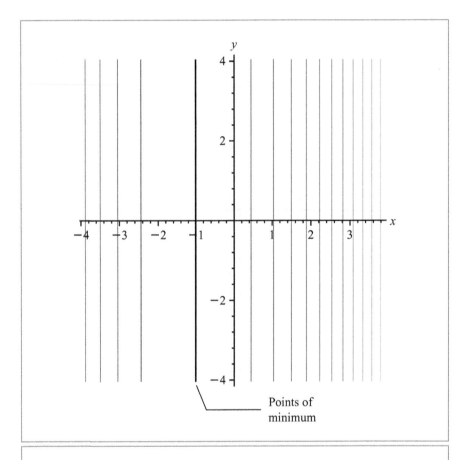

Figure 5.5: Quadratic function with an infinite number of local extrema.

which we can guarantee the ocurrence of the same phenomena that we observed in the examples (a) and (c) above, where the local minima that we computed were, in fact, global minima.

A function $f \colon \mathbb{R}^n \to \mathbb{R}$ is *convex* when

$$f(\lambda x_1 + (1 - \lambda)x_2) \le \lambda f(x_1) + (1 - \lambda)f(x_2)$$

for every $x_1, x_2 \in \mathbb{R}^n$ and $0 < \lambda < 1$. That is, the value of f at an interior point of a segment with extremes x_1 and x_2 is at most equal to the result obtained by linear interpolation between $f(x_1)$ and $f(x_2)$.

Theorem 6. *If $f: \mathbb{R}^n \to \mathbb{R}$ is convex and f has a local minimum at x_0, then f has a global minimum at x_0.*

Proof. Suppose that x_0 is not a global minimum point of f. Then, there exists x_1 such that $f(x_1) < f(x_0)$. Thus, for every point $\lambda x_0 + (1 - \lambda)x_1$ in the line segment $x_0 x_1$, we have

$$f(\lambda x_0 + (1 - \lambda)x_1) \leq \lambda f(x_0) + (1 - \lambda)f(x_1) < f(x_0).$$

Thus, f cannot have a local minimum at x_0. Therefore, if x_0 is a local minimum, then it is also necessarily a global minimum point.   □

It is not always easy to verify the convexity of a function using the definition. The following characterization, for functions of class C^2, is used in practice (see Luenberger, 1984).

Theorem 7. *A function $f: \mathbb{R}^n \to \mathbb{R}$, of class C^2, is convex if and only if the Hessian matrix $H(x) = \nabla^2 f(x)$ is nonnegative for every x (i.e., $d^\top H(x)d \geq 0$ for every x and every d).*

The examples given in the previous sections with quadratic functions are, therefore, specific cases of the results stated above: functions of type $f(x) = b^\top x + x^\top Hx$ are convex when H is nonnegative. Therefore, when they possess a local minimum at a given point, they also have a global minimum there.

5.2 LEAST SQUARES

Probably the most frequent optimization problem in applied mathematics is the one of adjusting a certain linear model to a data set in such a way that the error is as small as possible. The simplest case is the following: given a list of ordered pairs of observations (x_i, y_i), $i = 1, \ldots, n$, we must find numbers a and b in such a way that the straight line of equation $y = ax + b$ provides the best adjustment to the data set (see Figure 5.6).

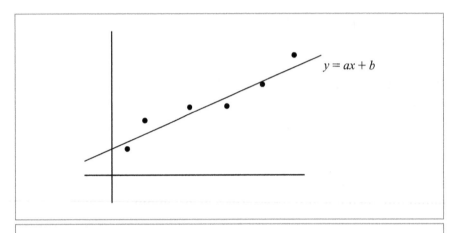

Figure 5.6: Linear regression.

This problem known as the *linear regression* problem can be given the following interpretation. We would like, if possible, to find values of a and b such that $y_i = ax_i + b$ for $i = 1,\ldots, n$. That is, we wish to solve the system of n equations with two variables:

$$X\beta = Y,$$

where

$$\beta = \begin{pmatrix} a \\ b \end{pmatrix}, \qquad X = \begin{pmatrix} x_1 & 1 \\ \vdots & \vdots \\ x_n & 1 \end{pmatrix}, \quad \text{and} \quad Y = \begin{pmatrix} y_1 \\ \vdots \\ y_n \end{pmatrix}.$$

Typically, n is greater than 2 and, in general, the above system has no solutions. That is, the vector Y is not in the subspace generated by the columns of X. Thus, we must settle for a vector β for which the error $|Y - X\beta|$ is as small as possible. That is, we need to minimize $|Y - X\beta|$ (or, equivalently, $|Y - X\beta|^2$) for $\beta \in \mathbb{R}^2$. But

$$|Y - X\beta|^2 = (Y - X\beta)^\top (Y - X\beta) = \beta^\top (X^\top X)\beta - 2Y^\top X\beta + Y^\top Y.$$

Therefore, the problem is a special case of the problem of minimizing a quadratic function in two variables, studied in Section 5.1.

Clearly, the matrix $X^T X$ is nonnegative. Moreover, since the space generated by the columns of $X^T X$ is equal to the space generated by the columns of X, the vector $Y^T X$ belongs to the subspace generated by the columns of $X^T X$. Therefore, f has a minimum at every point β_0 satisfying

$$(X^T X)\beta_0 = Y^T X$$

(the so-called *normal equations* of the least squares problem). If $X^T X$ is nonsingular (this occurs when the columns of X are linearly independent), then the system of normal equations has a unique solution, given by

$$\beta_0 = (X^T X)^{-1} Y^T X.$$

If $X^T X$ is nonsingular, then there exists an infinite number of vectors β_0 satisfying the equation (but only one β_0 of minimum length). In fact, we have the following general result.

Theorem 8. *Let A be an arbitrary matrix of order $m \times n$. For each $b \in \mathbb{R}^m$, there exists a unique vector x_0 of minimal length such that $|Ax_0 - b|$ is minimum. Moreover, the transformation that associates with each b the corresponding x_0 is linear. That is, there exists an $n \times m$ matrix A^+ such that $A^+ b = x_0$ for all b.*

Proof. We use geometric arguments in the proof (a more algebraic proof is provided in Strang, 1998). In the subspace $L = \{Ax | x \in \mathbb{R}^n\}$, there exists a unique vector y_0 whose distance to b is minimum (y_0 is the orthogonal projection of b on L) (see Figure 5.7). Moreover, there exists a unique vector x_0 of minimum length in the linear manifold $M(y_0) = \{x | Ax = y_0\}$ (x_0 is the orthogonal projection of the origin over $M(y_0)$). Thus, we have proved the existence and the uniqueness of x_0. Moreover, x_0 is obtained by the composition of two linear transformations: the one that maps $b \in \mathbb{R}^m$ to its orthogonal projection y_0 onto L and the one that maps y_0 to the projection of the origin onto $M(y_0)$. $\qquad\square$

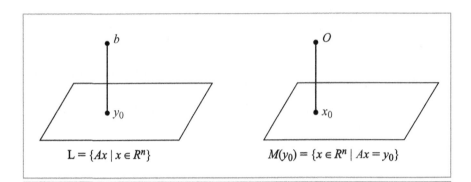

Figure 5.7: Geometric interpretation of the pseudoinverse.

The matrix A^+ in Proposition 8 is called the *pseudoinverse* of A. When A is $n \times n$ and is nonsingular, we have $\min|Ax_0 - b| = 0$, which occurs exactly at $x_0 = A^{-1}b$. Thus, the concept of pseudoinverse generalizes the concept of inverse matrix.

We should remark that the geometric interpretation given above can be related to the normal equations of the least squares problem. In fact, the normal equations can be written in the form $A^\top(Ax_0 - b) = 0$, which says that x_0 must be chosen in such a way that the vector $Ax_0 - b$ is orthogonal to the vector subspace $\{Ax | x \in \mathbb{R}^n\}$.

5.2.1 SOLVING LEAST SQUARES PROBLEMS

As we discussed above, it is natural to model the problem of "solving" an overconstrained system $Ax = b$ with more equations than variables (therefore, in general, without solutions) as the problem of minimizing the error $|Ax - b|$. Also, the resulting problem is equivalent to minimizing a quadratic function and therefore can be solved using the system of normal equations $(A^\top A)x_0 = A^\top b$.

This assertion, however, hides some difficulties with respect to the computation of x_0. It is very common, in this type of problem, that although the matrix $A^\top A$ is nonsingular, it is "almost" singular. As a result, the solution to normal equations using traditional methods (e.g., Gaussian elimination) is unstable from the numerical point of view (i.e., small perturbations in b might result in huge changes in x_0). One way to avoid this difficulty (or this problem) is to use specific methods for symmetric positive matrices, such as Cholesky's method (see Golub and Loan, 1983). Another alternative, which has the advantage of contemplating also the case where A^TA is singular, consists in using the technique of *singular value decomposition*. The following theorem, whose proof may be found in Golub and Loan (1983), can be interpreted as being an extension of the spectral theorem.

Theorem 9 [singular value decomposition]. *If A is a matrix of order $m \times n$ (with $m > n$), then there exist orthogonal matrices U and V of order, respectively, $m \times m$ and $n \times n$, such that $A = U \sum V$, where*

$$\sum = \begin{pmatrix} \sigma_1 & \cdots & 0 \\ \vdots & \ddots & \vdots \\ 0 & \cdots & \sigma_n \\ 0 & \cdots & 0 \\ \vdots & \ddots & \vdots \\ 0 & \cdots & 0 \end{pmatrix}$$

and $\sigma_1 \geq \sigma_2 \geq \cdots \geq \sigma_p > 0 = \sigma_{p+1} = \cdots = \sigma_n$.

There are efficient methods to compute such a decomposition (see, for instance, Press *et al.*, 1988). This decomposition deals very well with the fact that the columns of the matrix A might be "almost" linearly dependent. This reflects in the magnitude of the factors σ_i. Thus, we may fix a threshold T such that every $\sigma_i < T$ is made equal to 0.

The explanation of why this decomposition solves the problem of least squares resides in the fundamental fact that orthogonal transformations preserve distance. In fact, minimizing $|Ax - b|$ is equivalent to minimizing $|UAx - Ub|$. Also, $|y| = |Vy|$ for every y. Thus, by substituting $x = Vy$, the problem reduces to the one of obtaining y having minimum norm such that $|UAVy - Ub| = |\Sigma y - c|$ is minimum (where $c = Ub$). But because of the specific form of Σ, this problem is trivial. We must have $y_i = \frac{c_i}{\sigma_i}$, for $i = 1, \ldots, p$ and $y_i = 0$, para $i > p$. Once we have y, the desired solution is obtained by taking $x = Vy$.

5.3 ALGORITHMS

In the previous sections, we studied unconstrained optimization problems having quadratic objective function. For those problems, it is possible to apply the optimality conditions to obtain optimal solutions. They are obtained by solving, in an appropriate form, a system of linear equations. Moreover, quadratic problems have the outstanding quality that the optimal local solutions are also global solutions. For generic optimization problems, this does not occur: local extrema are not necessarily global. Even the determination of local extrema is complicated since the optimality conditions generate nonlinear equations, which makes it difficult to find the critical points. To find these points, one has to resort to iterative algorithms, almost always inspired in the behavior of quadratic functions: the basic idea is that, close to a local minimum, a function can be approximated, using Taylor's formula, by a quadratic function with the same point of minimum.

5.3.1 NEWTON'S METHOD

One of the better known methods to solve optimization problems is *Newton's method*. The fundamental idea of this method is to replace the function f, to be minimized or maximized, with a quadratic approximation based on the first and second derivatives of f. More precisely, let $f : \mathbb{R}^n \to \mathbb{R}$ be a function of class C^2 and x_0 a point sufficiently close to the desired solution. Taylor's formula of second order at x_0 provides a quadratic approximation to f:

$$\tilde{f}(x_0 + h) = f(x_0) + \nabla f(x_0)h + h^\top \nabla^2 f(x_0)h.$$

The approximation $\tilde{f}$ has minimum for $h = (\nabla^2 f(x_0))^{-1} \nabla f(x_0)^\top$ (provided that the Hessian matrix $\nabla^2 f(x_0)$ is nonnegative).

This step h generates a new approximation $x_1 = x_0 + h$ for the optimal solution, which is the starting point for the next iteration. Thus, Newton's method generates a sequence defined recursively by

$$x_{k+1} = x_k + (\nabla^2 f(x_k))^{-1} \nabla f(x_k)^\top.$$

It is interesting to observe that the terms of the sequence depend only on the derivatives of f and not on the values of the function f itself. In fact, Newton's method tries to find a stationary point of f, a point where the gradient is null. We have seen that this is a necessary condition, but it is not sufficient in order that f has a minimum at this point. If the sequence (x_k) converges (which is not guaranteed), it will converge to a point where $\nabla f = 0$. But this point is not necessarily a minimum of f. What is possible to guarantee is that if the point x_0 is chosen sufficiently close to a minimum x^* of f, then the sequence will converge to x^*, provided that f be sufficiently well behaved in the neighborhood of x^*. There are several criteria that define this good behavior. One such criterion is given by the theorem below, whose proof may be found in Luenberger (1984).

Theorem 10. *Let $f : \mathbb{R}^n \to \mathbb{R}$ be a function of class C^2. Suppose that f has a local minimum at x^*, and there exists a constant M and a neighborhood of x^* such that $|\nabla^2 f(x)| > M$ in this neighborhood. Then, there exists a neighborhood V of x^* such that if $x_0 \in V$, then the sequence defined by $x_{k+1} = x_k + (\nabla^2 f(x_k))^{-1} \nabla f(x_k)^\top$ converges to x^*. Moreover, the rate of convergence is quadratic, i.e., there exists c such that $|x_{k+1} - x^*| < c|x_k - x^*|^2$ for every k.*

Newton's method is eminently local and cannot be used as a general minimization method, starting from an arbitrary initial solution. Its use requires that other methods, such as the ones described below, be used to obtain a first approximation of the desired solution. Once this is done, Newton's method is an excellent refinement method, with an extremely fast convergence to a local extremum.

5.3.2 UNIDIMENSIONAL SEARCH ALGORITHMS

The fundamental idea of these algorithms is that of, at each iteration, reducing the n-dimensional problem of minimizing a function $f\colon \mathbb{R}^n \to \mathbb{R}$ to a simpler 1D optimization problem. To do that, a promising search direction at the current solution is chosen, and the 1D optimization problem obtained by restricting the domain of f to this direction is solved.

As an example, a more robust form of Newton's method can be obtained if x_{k+1} is the point that minimizes f along the line containing x_k and has the direction $(\nabla^2 f(x_k))^{-1}\nabla f(x_k)^\top$ provided by the method.

Another natural choice for the search direction is the direction where f has the greatest local descent rate, i.e., the direction along which the directional derivative of f has the smallest negative value. It is easy to prove that this direction is given by $-\nabla f$. Therefore, from a given point x_k, the next value is found by taking

$$x_{k+1} = \min f(x_k + \alpha_k d),$$

where $d = -\nabla f(x_k)$. This algorithm is known as the *steepest descent* method.

Depending on the problem, obtaining the exact solution to the above problem might not be practical. In fact, it is possible to prove the convergence of the method to a stationary point of f, even if the search of the next solution along d is inexact; it suffices for this that f be sufficiently reduced at each iteration (see Luenberger, 1984).

The steepest descent method has both advantage and disadvantage. The advantage is that under reasonable conditions, it has a guaranteed convergence to a point where $\nabla f(x^*) = 0$. The disadvantage is that this convergence may be very slow. This can be easily illustrated by observing its behavior for quadratic functions, represented in Figure 5.8 for $f(x, y) = x^2 + 4y^2$. Evidently, the minimum of f

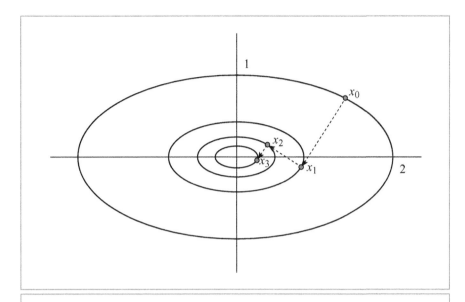

Figure 5.8: Gradient descent.

is attained at $(0,0)$. Starting from an arbitrary point, the sequence of solutions determines an infinite oscillation (in each step, the search is done in the direction of the normal to the level curve corresponding to the current solution, which never contains the origin).

It is possible to show that for quadratic functions of the form $f(x) = b^\mathsf{T} x + x^\mathsf{T} Hx$, with H positive, the sequence generated by the method converges to the minimum point x^* in such a way that

$$\frac{|x_{k+1} - x^*|}{|x_k - x^*|} < \frac{(A - a)^2}{(A + a)^2},$$

where a and A are, respectively, the smallest and the largest eigenvalues of H. If the ratio $\frac{A}{a}$ is large, then the gain in each iteration is small, and this leads to a slow convergence process.

5.3.3 CONJUGATE GRADIENT

The *conjugate gradient* algorithm also uses the idea, present in the steepest descent algorithm, of doing searches along promising directions, but it tries to choose these directions in such a way that the minimum point is reached after a finite number of steps (at least for quadratic functions).

Again, we use the 2D case to illustrate these ideas. There exists a case in which the steepest descent algorithm works in a finite number of steps for a quadratic function: when the level curves are circles, the normal to one of these curves at any point passes through the center. Therefore, the algorithm converges in a single iteration. When the level curves are ellipses, it is also possible to do the same, provided that the search direction be chosen not as a direction perpendicular to the tangent to the level curve but as a direction *conjugate* to it.

This can be understood in the following way. The level curves of a quadratic function $f(x) = x^\top (G^\top G)x$ (ellipses, in the general case) can be obtained from the level curves of $f(x) = x^\top x$ (which are circles) by applying the linear transformation $x \rightarrow Gx$. Conjugate directions of the ellipse are the images, by G, of the orthogonal directions (see Figure 5.9).

In general, if H is a positive matrix, two directions d_1 and d_2 are *conjugate* with respect to H when $d_1^\top H d_2 = 0$.

Given n mutually conjugate directions of $\mathbb{R}^n$, the minimum of a quadratic function $f: \mathbb{R}^n \rightarrow \mathbb{R}$ is computed by making n 1D searches, each along one of these directions.

For nonquadratic objective functions, the algorithm uses the Hessian matrix at x_0 to generate a set of n mutually conjugate directions. After searching along these directions, a better approximation to optimal solution is found, and a new set of directions is found.

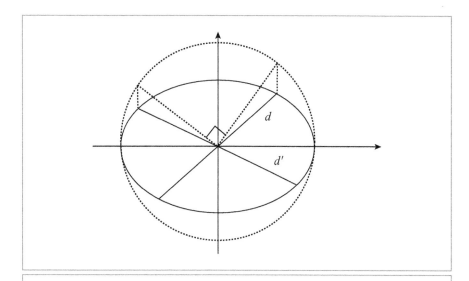

Figure 5.9: d and d' are conjugate directions.

The rate of convergence of the algorithm, under hyphothesis similar to those of Theorem 10, is quadratic, but with a constant greater than in Newton's method. However, because it incorporates information about the way the function decreases, the algorithm is more robust, and it is not necessary to provide an initial solution too close to the minimum point.

5.3.4 QUASI-NEWTON ALGORITHMS

The conjugate gradient algorithm, studied in the previous section, puts together good qualities of both Newton's method (rate of convergente) and steepest descent (robustness). These features are also found in the class of methods known as *quasi-Newton methods*.

These algorithms produce a sequence of solutions of the form

$$x_{k+1} = x_k - \alpha_k S_k \nabla f(x_k)^\top,$$

where S_k is a symmetric matrix and α_k is usually chosen in such a way as to minimize $f(x_{k+1})$. Both Newton's method (for which $S_k = (\nabla^2 f(x_k))^{-1}$) and steepest descent method (where $S_k = I$) are specific cases of this general form.

An important aspect that is considered in the construction of these algorithms is the fact that it may be impractical to compute the Hessian matrix of f (e.g., in the cases where f is not known analytically). Thus, the methods of this family usually construct approximations to the Hessian matrix based on the values of f and of ∇f in the last elements of the iteration sequence (see Luenberger, 1984, for details).

5.3.5 THE LEVENBERG–MARQUARDT ALGORITHM

A very common situation in applied mathematics in general and in computer vision in particular is parameter estimation. In this kind of problem, output data y depend on input data x through a function $g(\theta, x)$, where θ is a vector of unknown parameters to be estimated based on a sample of n pairs (x_i, y_i). Usually, the problem of finding the optimal parameter vector is set as the problem of minimizing the total squared error $\sum_{i=1}^{n} \left| g(\theta, x_i) - y_i \right|^2$.

In Section 5.2, we studied the special case where g is linear on θ, for which an analytical solution is available. For the general case of a nonlinear g, one has to resort to an iterative algorithm. Although a general-purpose optimization algorithm can be used, one can take advantage of the special structure of the problem (minimizing a sum of squares). The Levenberg–Marquardt (LM) algorithm is a variant of Newton's algorithm that exploits such structure.

The two main ideas behind the LM algorithm are

a) taking advantage of the special form of the objective function to compute the gradient and an approximation to the Hessian of the objective function

b) applying a strategy that finds a suitable combination of steepest descent and Newton's iteration.

The problem to be solved by the LM algorithm can be written as

$$\min f(\theta) = \frac{1}{2} \sum_{i=1}^{n} r_i(\theta)^2,$$

where $r_i(\theta)$ is the i-th residual $|g(\theta, x_i) - y_i|$ (the $\frac{1}{2}$ is just for convenience in expressing derivatives).

The gradient $G(\theta) = \nabla f(\theta)$ and the Hessian $H(\theta) = \nabla^2 f(\theta)$ of the objective function f can be expressed in terms of the residuals r_i and their Jacobian matrix $J\left(\text{with } J_{ij}(\theta) = \frac{\partial r_i}{\partial \theta_j}\right)$ as

$$G(\theta) = \sum_{i=1}^{n} r_i(\theta)\nabla r_i(x) = J(\theta)^T r(\theta)$$

$$H(\theta) = J(\theta)^T J(\theta) + \sum_{i=1}^{n} r_i(\theta)\nabla^2 r_i(\theta).$$

A key assumption about the LM algorithm is that the residual function is either small or near linear on θ (in such a way that $\nabla^2 r_i(\theta)$ is small). In either case, it is reasonable to approximate the Hessian H as

$$H(\theta) = J(\theta)^T J(\theta).$$

This assumption is reasonable when a good initial guess for θ is available. If this is not the case, the LM algorithm may perform badly compared to general-purpose optimization algorithms.

As mentioned above, the other key idea of the LM algorithm is the use of an adaptive update step, which provides a blend of Newton-type and steepest descent algorithms. The original update proposed by Levenberg (1944) was

$$\theta_{k+1} = \theta_k - (H(\theta_k) + \lambda I)^{-1} G(\theta_k), \tag{5.1}$$

where λ is a parameter that changes appropriately during the execution of the algorithm (as explained below). When λ is small, the update is similar to a Newton update; when λ is large, the algorithm behaves as a descent algorithm (with a step size that decreases with λ).

In the variant proposed by Marquardt (1963), the update is

$$\theta_{k+1} = \theta_k - (H(\theta_k) + \lambda \operatorname{diag}(H(\theta_k)))^{-1} G(\theta_k). \tag{5.2}$$

Using the diagonal of H instead of the identity allows one to take into account information about the curvature of H even when λ is large.

In both versions, the scaling parameter λ changes dynamically, according to whether a given update was successful or not at reducing the value of the objective function. If it was successful, the value of λ is decreased, moving closer to a (more efficient) Newton-type update. Otherwise, the algorithm adopts a more conservative strategy, increasing the value of λ, thus leaning toward a less efficient but safer descent strategy.

The entire algorithm can be described as follows:

1. Obtain a starting solution θ_0 (usually using some heuristic procedure); start with a moderate value of λ (for instance, $\lambda = 1$).
2. From a given solution θ_k, find θ_{k+1} according to the update given in 5.1 or 5.2.
3. If $f(\theta_{k+1}) < f(\theta_k)$, accept the update and decrease λ by a factor (10 is a common choice).
4. Otherwise, reject the update, increase λ by a factor (10, again, is a common choice), and try again an update.

Implementations of the LM algorithm are available at a number of sources (Press *et al.,* 1988 and the GNU and MINPACK scientific libraries).

5.4 CONSTRAINED OPTIMIZATION

In this section, we briefly discuss continuous optimization problems with constraints defined by equations or inequalities. That is, problems of the form

$$
\begin{aligned}
\min \ &f(x) \\
\text{subject to } &h_i(x) = 0, \quad i = 1, \ldots, m \\
&g_j(x) \le 0, \quad j = 1, \ldots, p,
\end{aligned}
$$

where every function is smooth (typically of class C^2).

5.4.1 OPTIMALITY CONDITIONS

The central idea of the first-order necessary conditions in the unconstrained case is that a local minimum occurs at points where every direction is stationary, i.e., those points where $\nabla f(x) \cdot d = 0$ for every d, which is equivalent to the condition $\nabla f(x) = 0$. In the proof, we argued that if f were not stationary along some direction d, we could obtain values smaller than $f(x)$ by moving along d or $-d$.

When the problem has constraints, there may be nonstationary directions at a local minimum point. Consider, for instance, the case where we wish to minimize $f(x)$ subject to the single constraint $h(x) = 0$. The feasible solutions to this problem are the points of a surface S of $\mathbb{R}^n$ (Figure 5.10). A point x_0 can be a local minimum of f restricted to S without having $\nabla f(x_0) = 0$. In fact, the component of $\nabla f(x_0)$ in the direction normal to the plane is irrelevant to the optimality of x_0 because, locally, x cannot move along this direction. Thus, we may have optimality, provided that the projection of $\nabla f(x_0)$ onto the tangent plane to S at x_0 is null (i.e., when we have $\nabla f(x_0) = \lambda \nabla h(x_0)$ for some real number λ).

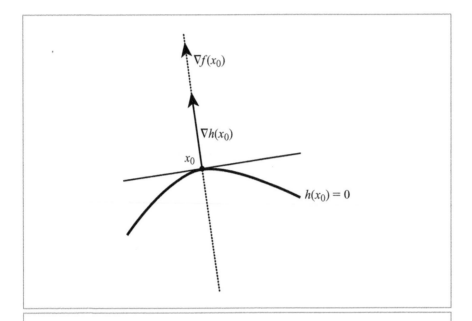

Figure 5.10: Equality constraint.

Suppose now that we have a single constraint of the form $g(x) \leq 0$. Again, it is possible to have optimality at x_0 with $\nabla f(x_0) \neq 0$. This can occur only if $g(x_0) = 0$, i.e., if x_0 is a point of the boundary surface of the region defined by $g(x_0) \leq 0$ (we say, in this case, that the inequality constraint is *active* at x_0). Moreover, $\nabla f(x_0)$ must be of the form $\lambda \nabla g(x_0)$. But now λ must be greater than or equal to zero. On the contrary, it would be possible to find solutions whose values are smaller than $f(x_0)$ by moving along the direction $-\nabla g(x_0)$ (see Figure 5.11).

These ideas constitute the basis for the demonstration of the theorem that follows, which provides first-order necessary conditions for the general case of m equality constraints and p equality constraints. We say that x_0 is a *regular point* for these constraints when the vectors $\nabla f_i(x_0)$ ($i = 1, \ldots, n$) and $\nabla g_j(x_0)$ (j such that $g_j(x_0) = 0$) are linearly independent.

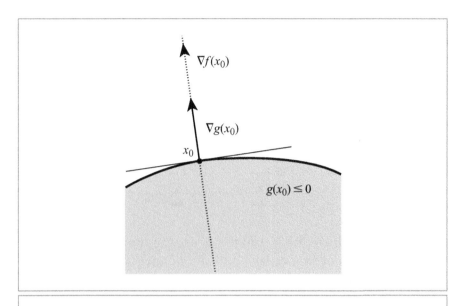

Figure 5.11: Inequality constraint.

Theorem 11 [Kuhn-Tucker condition]. *Let x_0 be a relative minimum point to the problem*

$$\min \; f(x)$$
$$\text{subject to} \; h_i(x) = 0, \quad i = 1, \ldots, m$$
$$g_j(x) \leq 0, \quad j = 1, \ldots, p,$$

and suppose that x_0 is a regular point for these constraints. Then, there exists a vector $\lambda \in \mathbb{R}^m$ and a vector $\mu \in \mathbb{R}^p$, with $\mu \geq 0$, such that

$$\nabla f(x_0) + \sum_{i=1}^{n} \lambda_i \nabla h_i(x_0) + \sum_{j=1}^{p} \mu_j \nabla g_j(x_0) = 0$$
$$\sum_{j=1}^{p} \mu_j g_j(x_0) = 0.$$

The theorem above generalizes the classical theory of Lagrange multipliers for the case where inequality constraints are present.

Also, there are theorems that provide second-order necessary and sufficient conditions to ensure that a given point x_0 is a local minimum. In the same way as in the unconstrained case, such conditions refer to the nonnegativeness (or the positiveness) of a certain Hessian matrix.

In the constrained case, those conditions apply to the Hessian of the *Lagrangian function*, given by

$$L(x) = f(x) + \sum_{i=1}^{n} \lambda_i h_i(x) + \sum_{j=1}^{p} \mu_j g_j(x_0)$$

at the point x_0 (where the λ_i and the μ_j are the multipliers provided by the conditions of first order), and the nonnegativeness (or positiveness) is imposed only for directions of the tangent space, at x_0, to the surface determined by the active constraints at this point (see Luenberger, 1984, for details).

5.4.2 LEAST SQUARES WITH LINEAR CONSTRAINTS

We use the least squares problem with linear constraints to exemplify the use of the above theorem. Let us consider the following problem:

$$\min |Ax - b|$$
$$\text{subject to } Cx = D,$$

where A is $m \times n$ (with $m > n$) and C is $p \times n$ (with $p < n$).

In this problem, there exists a set of equations that must be satisfied exactly and another set of equations that will be satisfied in the best possible way.

Minimizing $|Ax - b|$ is equivalent to minimizing $|Ax - b|^2$, i.e., the problem is equivalent to

$$\min \ x^\top (A^\top A)x - 2b^\top Ax + b^\top b$$
$$\text{subject to } \ Cx = D.$$

In order that x_0 be a relative minimum point, it must satisfy the first-order conditions. In this case, where there are only equality constraints, there must exist a vector $\lambda = (\lambda_1, \lambda_2, \ldots, \lambda_p)^\top$ such that

$$(A^\top A)x_0 - A^\top b + C^\top \lambda = 0.$$

Since x_0 must also satisfy the equality constraints, we obtain the system

$$\begin{pmatrix} A^\top A & C^\top \\ C & 0 \end{pmatrix} \begin{pmatrix} x \\ \lambda \end{pmatrix} = \begin{pmatrix} A^\top b \\ d \end{pmatrix}.$$

It is possible to prove that, provided that the rows of C are linearly independent, the above system always has at least one solution x_0 obtained by finding the closest point to b among the points of the linear manifold $\{Ax | Cx = d\}$. As it happens in the unconstrained case, such solutions are, in fact, global minima of the least squares problem.

In order to effectively solve the above system, it is advisable, again, to use the singular value decomposition of $A^\top A$ to avoid the numerical instability that originates from the possible quasi-singularity of $A^\top A$.

5.4.3 ALGORITHMS

There is a variety of algorithms for continuous optimization problems with constraints. Some of these algorithms are general, and they are capable of dealing with any combination of constraints by equalities or inequalities. Others try to exploit the nature of the constraints—equality or inequality—and properties of the functions that define them (and those of the objective function).

In what follows, we describe two classes of algorithms for optimization with constraints. As we will see, these algorithms exploit the following ideas:

- reduce the constrained problem to a problem without constraints (or a sequence of problems of this kind) in such a way as to allow the use of the algorithms previously described.

- adapt an algorithm for unconstrained optimization to take constraints under consideration.

Penalty and Barrier Methods

Methods in this class exploit the first idea above by modifying the objective function in such a way as to take constraints into account.

Penalty methods are more general than barrier methods and work with both equality and inequality constraints. Moreover, they do not require (in contrast to other methods), as we know, a feasible initial solution. In fact, penalty methods constitute a good method to find such a feasible initial solution.

The basic idea of the method consists in, for each constraint, introducing in the objective function a *penalty* term that discourages the violation of this constraint. For a constraint of the form $h(x) = 0$, we introduce a penalty of the form $ch(x)^2$, and for a constraint of the form $g(x) \leq 0$, a penalty of the form $c \max\{0, g(x)^2\}$ is appropriate. In this way, for the constrained problem

$$\begin{aligned}
\min \ & f(x) \\
\text{subject to} \ & h_i(x) = 0, \quad i = 1, \ldots, m \\
& g_j(x) \leq 0, \quad j = 1, \ldots, p,
\end{aligned} \tag{5.3}$$

we obtain an unconstrained problem of the form

$$\min q_c(x) = f(x) + c \left(\sum_{i=1}^{m} h(x)^2 + \sum_{j=1}^{p} \max(0, g(x)^2) \right).$$

The simple introduction of penalties in the objective function does not guarantee that the resulting problem has an optimal solution satisfying the constraints. The strategy of the penalty method is to solve a sequence of problems as the one above, with c taking the values of an increasing sequence (c_k) such that $c_k \to \infty$.

The following theorem guarantees that this strategy works.

Theorem 12. *Let (c_k) be an increasing sequence such that $c_k \to \infty$. For each k, let $x_k = \arg\min q_{c_k}(x)$. Then, every adherence point of*[1] *(x_k) is a solution to 5.3.*

Barrier methods are quite similar to penalty methods. The difference is that they operate in the interior of the solution set. For this reason, they are typically used for problems with inequality constraints.

The idea is to add for each inequality $g(x) \le 0$ a term in the objective function that avoids the solutions to get too close to the surface $g(x) = 0$. The introduction of this term, which can be, for instance, of the form $-\frac{c}{g(x)}$, works as a *barrier* that avoids this approximation. The resulting problem is given by

$$\min \ q_c(x) = f(x) - c \sum_{j=1}^{p} \frac{1}{g(x)}$$
$$\text{subject to } \ g_j(x) \le 0, \qquad j = 1, \dots, m.$$

In spite of the fact that the constraints are still present in the formulation of the problem, they may be ignored if the optimization method used does searches starting from a solution that belongs to the interior of the solution set S: the barrier introduced in the objective function automatically ensures that the solutions stay in the interior of S.

Again, it is not enough to solve the above problem; it is necessary to consider a sequence (c_k) of values of c. Here, however, we must take this sequence in such a way that it is decreasing and has limit 0 (this is equivalent to removing, progressively, the barriers in such a way as to include all of the set S). As in the case of penalty methods, the sequence of solutions (x_k) generated in the process has the property that its adherence points are solutions to the original problem.

---

1   An *adherence point* of a sequence is the limit of any of its subsequences.

Penalty methods and barrier methods have the virtue of being very general methods and relatively simple to be implemented because they use optimization methods without constraints. Nevertheless, their convergence is typically slow.

Projected Gradient Methods

Here, we consider only problems with equality constraints:

$$
\begin{aligned}
\min\ & f(x) \\
\text{subject to}\ & h_i(x) = 0, \quad i = 1, \ldots, m. \\
& (x \in \mathbb{R}^n)
\end{aligned}
\tag{5.4}
$$

In principle, the solution set S is a manifold of dimension $n - m$, and ideally, the above problem could be reduced to an unconstrained problem of the same dimension.

One of the few cases where this can be done in practice is when the constraints are linear. In fact, if the constraints are of the form $Cx = d$ (where C is a matrix $m \times n$ of rank m), S is a linear manifold of dimension n, and it can be expressed in the form

$$
S = \{Py + q | y \in \mathbb{R}^{n-m}\}.
$$

We obtain, therefore, the equivalent unconstrained problem of minimizing $g(y) = f(Py + q)$. The derivative of the new objective function may, naturally, be obtained from that of f by using the chain rule.

The above ideas are useful even when the constraints are not linear. Let x_0 be a feasible solution to 5.4, and consider the plane H, tangent to S at x_0, which can be defined by

$$
H = \{Py + x_0 | y \in \mathbb{R}^{n-m}\},
$$

where the columns of P constitute a basis for the orthogonal complement of the subspace generated by $((\nabla h_1(x_0))^{\mathsf{T}}, \ldots (\nabla h_m(x_0))^{\mathsf{T}})$.

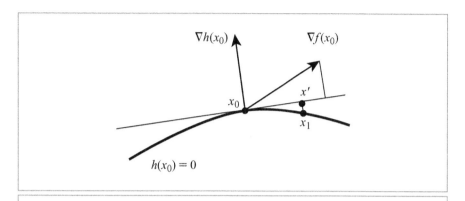

Figure 5.12: Projected gradient algorithm.

The plane H can be interpreted as an approximation to S. Moreover, the problem of minimizing f on H can be reduced to an unconstrained problem, as seen above. The steepest descent algorithm, applied to this problem, performs a search along the direction given by the projection, onto H, of the gradient vector in x_0. The new solution x', thus, obtained does not satisfy, in general, the constraints $h_i(x) = 0$ but is near the manifold S. From x' a new feasible solution x_1 can be computed (e.g., using a 1D search in the direction normal to S, as illustrated in Figure 5.12). The process is then iterated, resulting in the *projected gradient* algorithm, which has convergence properties similar to those of the steepest descent for the unconstrained case.

Other unconstrained optimization algorithms (e.g., the conjugate gradient method) can be similarly adapted to operate on the tangent space.

5.5 LINEAR PROGRAMMING

Now we study a class of optimization problems called *linear programs*, which have, simultaneously, characteristics of continuous and combinatorial optimizations. A linear program is a problem of the form

$$\min \sum_{j=1}^{n} c_j x_j$$

$$\text{subject to } \sum_{j=1}^{n} a_{ij} x_j \leq b_i, \quad i = 1, \ldots m.$$

That is, linear programs ask for minimizing a linear function subject to linear inequalities. In fact, the constraints can be constituted by any combination of linear equalities and inequalites: it is always possible to convert such a problem into another equivalent one of the above form. As an example, an equation of the form

$$\sum_{j=1}^{n} a_j x_j = b$$

can be replaced with the pair of inequalities

$$\sum_{j=1}^{n} a_j x_j \leq b;$$

$$\sum_{j=1}^{n} a_j x_j \geq b.$$

Linear programs are clearly continuous optimization programs with inequality constraints and can be solved by iterative algorithms such as those described in Section 5.4. In fact, algorithms of this form, specially formulated for linear programs, produce some of the most efficient methods to solve them, the so-called *interior point methods*.

Nevertheless, the first successful approach to linear programs—making it possible to solve problems having hundreds of thousands of variables—was combinatorial in nature. This work was done, in the '50s, by Dantzig, who introduced perhaps the most famous optimization algorithm—the *simplex method.*

In order to use the simplex method, it is more interesting to write a linear program in the *canonical form*:

$$\max \sum c_j x_j$$
$$\text{subject to } \sum a_{ij}x_j = b_i, \quad i = 1, \ldots, m$$
$$x_j \geq 0$$

or, using matrix notation,

$$\max cx$$
$$\text{subject to } Ax = b$$
$$x \geq 0,$$

where A is an $m \times n$ matrix.

In general, we suppose that the rank of A is equal to m (i.e., the rows of A are linearly independent). In this case, $\{x|Ax = b\}$ is a linear manifold (i.e., the translated vector subspace) of dimension $n - m$.

The fact that linear programs have an important combinatorial structure follows from the fact that the solution set $S = \{x|Ax = b, x \geq 0\}$ is a *convex polyhedra* and that all relevant information for the optimization of a linear function is concentrated on the vertices of the polyhedra.

We say that $\bar{x}$ is a vertex of S when $\bar{x}$ is not an interior point of any line segment with extremes in S. That is, $\bar{x} = \lambda x_1 + (1 - \lambda)x_2$ implies that $\bar{x} = x_1 = x_2$, for $x_1, x_2 \in S$, and $0 < \lambda < 1$.

The importance of vertices in the solution to linear programs is given by the following theorem.

Theorem 13. *If a linear program has an optimal solution, then it has a vertex that is an optimal solution* (see Figure 5.13).

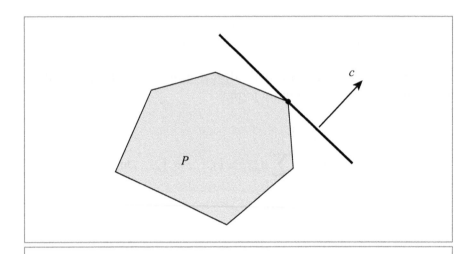

Figure 5.13: Optimal solutions to linear programs.

The geometric notion of vertex has an algebraic equivalent. We say that $\bar{x}$ is a *basic solution* to $S = \{x|Ax = b\}$, when

1. $A\bar{x} = b$
2. the columns j of A such that $\bar{x}_j > 0$ are linearly independent (they may, therefore, the set of such columns can be extended to a basis of the space generated by the columns of A).

Theorem 14. *$\bar{x}$ is a vertex if and only if $\bar{x}$ is a feasible basic solution.*

A consequence of the above theorems is that in order to solve a linear program, it suffices to examine each subset of m columns of A. We have, therefore, an naive algorithm to solve a linear program, which consists of the following steps:

1. Select m columns of A and verify if they constitute a basis.
2. Compute the solution corresponding to the previous step by attributing the value 0 to the other variables.
3. Verify if the solution obtained is feasible.

This sequence of steps must be repeated for every possible basis, and, among them, the one that produces the greatest value is the optimal solution. Note that there are $\binom{n}{m}$ possible bases, i.e., approximately n^m alternatives to be compared.

Thus, this algorithm, although finite, is clearly inefficient, and it is not useful to solve real problems of linear programming. In what follows, we describe the simplex method, which also restricts attention to the basic points. However, it uses the adjacency structure of the polyhedra vertices, producing a very efficient algorithm for practical problems.

5.5.1 SIMPLEX ALGORITHM FOR LINEAR PROGRAMS

The simplex algorithm generates a sequence of vertices of the solution set $S = \{x | Ax = b\}$, always decreasing the value of $f(x) = cx$, until it finds the minimum value or obtains a proof the the problem is unbounded.

The algorithm consists in executing, repeatedly, the following steps:

1. Verify if the current basic solution $\bar{x}$ is optimal. (For this, we just have to verify if $f(x) \geq f(\bar{x})$ for all of the vertices adjacent to $\bar{x}$. If this occurs, $\bar{x}$ is *optimal*. If not, we execute the following step.)
2. Move along an edge on which the objective function decreases until it hits a new vertex. (If no vertex is found when moving over an edge where f is decreasing, the problem is *unbounded*.)

As we have already seen, each basic solution is associated with a set of linearly independent columns of A. The variables corresponding to these columns are called *basic variables* for the solution.

The simplex method maintains an algebraic representation of the system $Ax = b$, where the nonbasic variables are expressed in terms of basic variables. With this representation, it is simple to execute, at each step, the optimality test and the passage to an adjacent vertex. Once this is done, it is necessary to update the system representation in such a way that the nonbasic variables are again expressed in terms of the basic variables. This updating, called *pivoting*, is executed using elementary operations on the equations of the system $Ax = b$ (see Chvatal, 1983, for details).

The simplex algorithm needs an initial feasible basic solution, which is not always obvious. However, it is interesting to remark that such a solution may be found by solving another linear program (using again the simplex method) for which a feasible basic solution is immediately available. For example, to obtain a feasible basic solution to $S = \{x | Ax = b\}$, we may solve the problem below, in which we introduce *artificial variables* x_a:

$$\min 1 \cdot x_a \quad (\text{i.e., } \sum xa_i)$$
$$Ax + Ix_a = b$$
$$x, x_a \geq 0.$$

This second problem has an obvious feasible basic solution given by $x = 0$ and $x_a = (1, 1, \ldots, 1)$. Moreover, it always has an optimal solution because 0 is a lower bound of the objective function. If the value of this optimal solution is 0, all the artificial variables x_a have value 0 and therefore the nonartificial variables provide a feasible solution to the original problem. Otherwise, the original problem does not have feasible solutions.

5.5.2 THE COMPLEXITY OF THE SIMPLEX METHOD

In the worst case, the simplex method may demand an exponential number of iterations to solve a linear program. In practice, however, the number of iterations is linear in m. However, the algebraic effort in each step has complexity $O(mn)$, which makes the method practical even for problems with a huge number of variables (large-scale problems).

The searching for guaranteed polynomial algorithms for linear programs was, during some decades, one of the most important open problems in optimization. The question was answered in a satisfactory way in the late '70s, when Khachian (1979) proved that the so-called *ellipsoid algorithm*, initially developed to solve nonlinear problems, could be used to solve linear programs in polynomial time. Subsequently, other algorithms were discovered that gave the same result. The most famous among them is *Karmarkar's algorithm* (Karmarkar, 1984). These algorithms instead of moving along the boundary of the solution set polyhedra $S = \{x | Ax = b\}$ produce a sequence of points in the interior of S that converge to an optimal solution.

The linear programs of greatest interest for computer graphics are, in general, of dimension 2 or 3. It is interesting to observe that such problems may be solved in linear time (Meggido, 1984). In fact, there exist linear algorithms to solve linear programs with any number of variables, provided that this number is assumed constant. However, such algorithms are practical only for low dimensions (and even in these cases, in practice the simplex method presents a comparable performance).

5.6 APPLICATIONS IN GRAPHICS

In this section, we discuss problems in computer graphics, which can be posed as continuous optimization problems. The list of problems we cover is

- camera calibration
- registration of an image sequence
- color correction of an image sequence
- real-time walk-through.

5.6.1 CAMERA CALIBRATION

Problems involving the estimation of parameters are naturally posed as a continuous optimization problems, where the objective is to determine the set of parameters that minimize the adjustment error. One important example in computer graphics is the problem of calibrating a camera, i.e., estimating the intrinsic (characteristics of its optical system) and extrinsic (position and orientation) parameters.

The simplest camera model is the one that has no optical system (no lens): it corresponds to the rudimentary model of the *pinhole camera* where light enters through a small hole in a box, projecting itself onto a plane, as illustrated in Figure 5.14. The image produced is simply a perspective projection of the 3D scene. This simple camera model is the basic model used in computer graphics to define the virtual camera for producing synthetic images. For this reason, applications that involve capture and synthesis of images commonly use this model.

The distance between the optical center of the camera and the plane of projection is called the *focal distance* and is denoted by *f*. This camera model has no other intrinsic parameter other than the focal distance, and its perspective projection is characterized by this parameter and a set of extrinsic parameters. In order to identify these parameters, it is convenient to consider three reference systems (Figure 5.15): a system *Oxyz* of *world coordinates*, used to reference the position of points in space; a system *C'x'y'z'* of *camera*

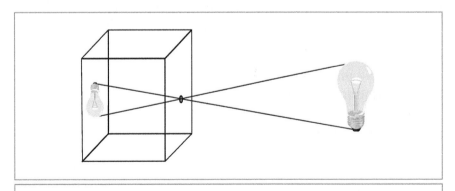

Figure 5.14: The pinhole camera.

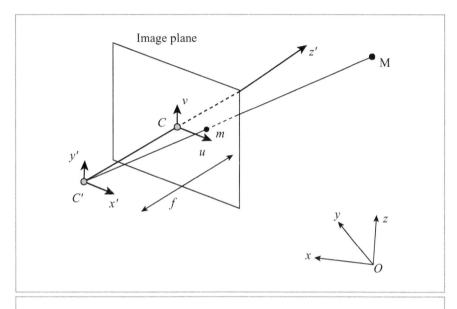

Figure 5.15: Camera coordinate systems.

coordinates, with origin at the optical center and axes aligned with the camera principal directions (i.e., the directions of the image boundary and the normal to the image plane); and a system of 2D coordinates, called *image coordinate system Cuv*, obtained by projecting the camera reference system on the plane of the virtual screen.

The camera extrinsic parameters may then be expressed using the the rotation matrix R and the translation t that expresses the coordinate transformation between the world coordinate system and the camera coordinate system. It is important to observe that although the matrix R has nine elements, it has only 3 degrees of freedom because it is an orthogonal matrix.

Thus, for a given camera with parameters f, t, and R, the function that provides the image $m = (u, v)$ of a point $M = (x, y, z)$ of space can be computed in two steps.

- In the camera reference system, the coordinates of M are given by

$$\begin{pmatrix} x' \\ y' \\ z' \end{pmatrix} = R \begin{pmatrix} x \\ y \\ z \end{pmatrix} + t.$$

- In the camera reference system, the image of a point (x', y', z') is given by

$$u = \frac{fx'}{z'}, \quad v = \frac{fy'}{z'}.$$

Composing the two transformations, we obtain the following equation:

$$\begin{pmatrix} u \\ v \\ w \end{pmatrix} = \begin{pmatrix} \mathbf{r}_1 & t_x \\ \mathbf{r}_2 & t_y \\ f\mathbf{r}_3 & ft_z \end{pmatrix} \begin{pmatrix} x \\ y \\ z \\ 1 \end{pmatrix},$$

which provides, in homogeneous coordinates, the image (u, v, w) of a point $(x, y, z, 1)$ of the space by a camera with parameters f, t, and R. In the above equation, $\mathbf{r}_1$, $\mathbf{r}_2$, and $\mathbf{r}_3$ are the vectors that constitute the rows of the rotation matrix R. These equations are used, for example, to generate the projection and a synthetic image of a 3D scene.

Suppose now that we are given the image of a scene and that we need to add to this image some additional synthetic elements. For this, it is necessary to generate a synthetic camera with the same parameter values of the camera that captured the scene. If these parameters are not known, they must be estimated based on information from the image.

Suppose that there exist n points in the image whose spatial coordinates $m_i = (x_i, y_i, z_i)$ $(i = 1, \ldots, n)$ are known. By associating these coordinates with the corresponding coordinates (u_i, v_i) on the image, we obtain $2n$ equations whose variables are the camera parameters. Therefore, the computation of the parameters consists in solving this system of equations.

However, in general, it is not possible (neither desired) to solve this system exactly. Typically, we use a large number of points in order to calibrate the camera because using a small number of samples leads, in general, to considerable estimation errors. Thus, the system of equations generated by associating points in space with corresponding points in the image has more equations than variables and, in general, does not have a solution. That is, for any set of parameters, there will be discrepancies between the observed image (u_i, v_i) of a point (x_i, y_i, z_i) and the image $(\tilde{u}_i, \tilde{v}_i)$ obtained with these parameters. The set of parameters to be adopted must then be chosen in such a way that these discrepancies are as small as possible. It is very common to use the quadratic error as a measure of the discrepancy

$$\sum_{i=1}^{n} (u_i - \tilde{u}_i)^2 + (v_i - \tilde{v}_i)^2.$$

Thus, the camera calibration problem can be posed as

$$\min_{t, f, R} \sum_{i=1}^{n} (u_i - \tilde{u}_i)^2 + (v_i - \tilde{v}_i)^2,$$

where R must be an orthogonal matrix, i.e., it must satisfy the equations given by $RR^\top = I$. In order that the problem can be treated as unconstrained, it is necessary to express R in terms of only three parameters (e.g., using Euler angles or unit quaternions (Watt, 2000)).

The LM algorithm, described in Section 5.3.5, is commonly used to solve this optimization problem. However, it requires a good starting solution, and specific iterative methods have been devised for this purpose. The most popular is *Tsai's algorithm* (Tsai, 1987), which also considers the presence of radial deformations caused by an optical system.

In some applications, it might not be necessary to recover the camera parameters. It is sufficient to find a matrix Q such that $(u\,v\,w)^\top = Q (x\,y\,z\,1)^\top$. This occurs, for example, when we only need

to associate other points of the space with its projections, without
building a synthetic camera. The problem turns out to be the one of
finding

$$Q = \begin{pmatrix} \mathbf{q}_1 & q_{14} \\ \mathbf{q}_2 & q_{24} \\ \mathbf{q}_3 & q_{34} \end{pmatrix}$$

in such a way to minimize $\sum_{i=1}^{n} (u_i - \tilde{u}_i)^2 + (v_i - \tilde{v}_i)^2$, where

$$\tilde{u}_i = \frac{\mathbf{q}_1 m_i + q_{14}}{\mathbf{q}_3 m_i + q_{34}} \quad \text{and} \quad \tilde{v}_i = \frac{\mathbf{q}_2 m_i + q_{24}}{\mathbf{q}_3 m_i + q_{34}}.$$

This problem is still nonlinear, but an acceptable solution can be
obtained by solving the least squares problem obtained by multiply-
ing each term of the above sum by the square of the common denom-
inator of $\tilde{u}_i$ and $\tilde{v}_i$. In this way, we can reduce it to the problem of
minimizing

$$\sum_{i=1}^{n} (\mathbf{q}_1 m_i + q_{14} - u_i(\mathbf{q}_3 m_i + q_{34}))^2 + (\mathbf{q}_2 m_i + q_{24} - v_i(\mathbf{q}_3 m_i + q_{34}))^2.$$

It is still necessary to add a normalization constant because the
above expression is obviously minimized when $Q = 0$. We may adopt,
for instance, $|q_3| = 1$, which leads us to an eigenvector problem,
as the one discussed by Faugeras (1993) or $q_3 \bar{m} = 1$, where $\bar{m} = \frac{1}{n} \sum_{i=1}^{n} m_i$ (which can be solved using the techniques discussed in
Section 5.2).

The matrix Q obtained by the above process does not correspond, in
general, to a camera. Nevertheless, it can be used as a starting point to
obtain a camera if this is necessary for the application (see Carvalho
et al., 1987).

5.6.2 REGISTRATION AND COLOR CORRECTION FOR A SEQUENCE OF IMAGES

An area of computer graphics that has received a great attention recently is *image-based rendering*. The basic idea of this area is to reconstruct a 3D scene model from a finite number of images of the scene.

Several of these applications make this visualization from a single *panoramic image* of the scene, i.e., from an image taken using an angle of 360° around a viewpoint as shown in Figure 5.16. An example of an application that uses visualization based on panoramic images is the *Visorama*, which uses a special device to simulate an observer who observes the scene using a binocular (see Matos *et al.*, 1997).

Although there are special photographic cameras capable of capturing panoramic images, it is possible to construct such images by matching a set of conventional photographs taken from the same viewpoint and covering an angle of 360° of the scene. In order to enable the correct registration in the matching process, consecutive photographs must have an overlapping area, as illustrated in Figure 5.17.

The construction of the panoramic image requires the solution to two problems, and both can be tackled using optimization techniques.

Figure 5.16: A panoramic image.

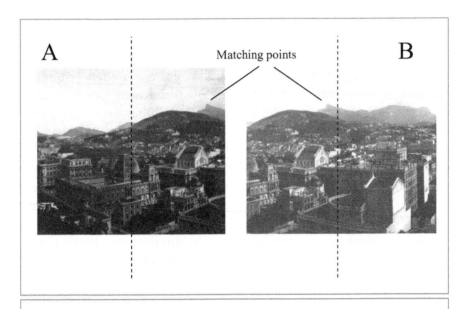

Figure 5.17: Image sequence with overlapping.

The first of these problems is geometric registration. This problem has elements that are common to the camera calibration problem studied in Section 5.6.1. In order to construct the panoramic image, it is necessary to recover the camera transformations associated with each photograph. This can be done, for instance, by identifying common points in the overlapping areas of the photos (as indicated in Figure 5.17), allowing us to relate each photo to its neighbors. By taking into consideration that the position of the camera is the same for both photographs, points in them are related by a transformation of the form

$$
\begin{pmatrix} x' \\ y' \\ w' \end{pmatrix} = \begin{pmatrix} 1 & 0 & 0 \\ 0 & 1 & 0 \\ 0 & 0 & 1/f_2 \end{pmatrix} R \begin{pmatrix} 1 & 0 & 0 \\ 0 & 1 & 0 \\ 0 & 0 & f_1 \end{pmatrix} \begin{pmatrix} x \\ y \\ w \end{pmatrix}, \qquad (5.5)
$$

where f_1 and f_2 are the focal distances of the cameras and R is the rotation matrix that expresses the position of the second camera with respect to the first one. These values must be computed in such a way as to minimize the error in the above transformation.

Thus, it is possible from the first photo to recover successively the focal distance and the orientation (relative to the first camera) of the cameras used in the other photos of the sequence. However, better results are obtained by doing, afterward, a joint calibration of all cameras (this process is called *bundle adjustment*). After getting the cameras, the panoramic image can be constructed by applying the appropriate transformation to each photo.

There also exist methods capable of executing the registration process automatically, without being necessary to identify manually the corresponding points in the superposition regions. Such methods operate directly with the images and try to find the camera parameters that maximize the similarity among them (see Szeliski and Shum, 1997, and McMillan and Bishop, 1995).

The second problem related to "stitching" the photographs to produce the panoramic image is the color compatibility between the photos. Although the photos are obtained by the same camera, from the same point, and almost at the same time, several factors contribute to the generation of color differences among the corresponding objects of the image sequence. In Figure 5.17, for instance, we observe that the overlapping region of image A is darker than the corresponding region in the image B.

The solution to this problem is described by Pham and Pringle (1995) and consists in finding a transformation that corrects the colors in image A to make them compatible with those in image B. To compute such a transformation, we use the colors present in the superpositon area of A and B. We find an association between the levels (r, g, b) in the two images of the form

$$r_B = p_r(r_A, g_A, b_A)$$
$$g_B = p_g(r_A, g_A, b_A)$$
$$b_B = p_b(r_A, g_A, b_A),$$

where p_r, p_g, and p_b are polynomials.

As an example, we use second-degree polynomials of the form

$$p(r_A, g_A, b_A) = c_{000} + c_{100}r_A + c_{010}g_A + c_{001}b_A + c_{200}r_A^2 + c_{020}g_A^2$$
$$+ c_{002}b_A^2 + c_{110}r_Ag_A + c_{101}r_Ab_A + c_{011}g_Ab_A,$$

although Pham and Pringle (1995) report better results using third-degree polynomials. The coefficients of each of the polynomials p_r, p_g, and p_b are then computed in such a way as to minimize the quadratic error resulting from the adjustment. That is, we solve a problem of the form $\min|X\beta - Y|$, where

- β is a 10×3 matrix containing the parameters to be determined (a column for each polynomial)
- Y is an $n \times 3$ matrix where each row contains a color (r_B, g_B, b_B) of the image B
- X is an $n \times 10$ matrix where each row is of the form

$$\left(1 \ r_A \ g_A \ b_A \ r_A^2 \ g_A^2 \ b_A^2 \ r_Ag_A \ r_Ab_A \ g_Ab_A\right)$$

and corresponds to a color of B.

This problem can be solved using the methods discussed in Section 5.2.

5.6.3 VISIBILITY FOR REAL-TIME WALK-THROUGH

Virtual environment real-time walk-through demands the use of procedures that guarantee that the amount of data sent to the visualization pipeline be limited in order to maintain an adequate visualization rate (or frame rate).

We describe here a technique suitable for *walk-through* in building environments. Although the modeling of large-scale building environments demands a huge number of objects, typically only a small number can be visible simultaneously because most of them are hidden by obstacles (such as the walls of the building). The strategies described below can be used to accelerate the visualization of such structures.

The first consists in organizing the polygons that compose the scene in a hierarchical structure called *binary space partition tree*, or *BSP tree* The use of this structure allows that the polygons located behind the observer be eliminated in the visualization process. Moreover, this structure provides a depth ordering for the other polygons, which allows us to avoid the use of the *z-buffer* or other strategies for removing hidden faces (see Fuchs *et al.,* 1980). Figure 5.18 illustrates this idea. The plane is initially divided into two half-planes by the line r_1; each of these two half-planes is in turn divided into two regions by lines r_2 and r_3, and so on. In the end of the process, we have a tree of subdivisions where each leaf corresponds to one of the 16 regions (or cells) delimited in the figure.

The use of the BSP tree structure allows us to reduce the number of objects for visualization. If the observer is in cell 1, looking in the indicated direction, the use of the BSP tree structure guarantees that only the marked walls need to be sent to the visualization pipeline.

In the case of huge environments, this is not sufficient. As an example, if the observer is in cell 5, looking in the indicated direction, only the walls to the left would be excluded in the visualization using the BSP tree analysis, although other walls are also invisible from any observer's position in this cell.

To cope with this situation, Teller and Sequin (1991) proposes the adoption of an additional structure capable of storing visibility information, relative to the different cells determined by the BSP tree. This additional information is stored in a *cell–cell visibility graph*. In this graph, there exists an edge connecting two cells when one of them is visible from the other. That is, when there exist points interior to these two cells such that the segment that connects them does not cross any wall. From this graph, the polygons corresponding to one cell are rendered only if this cell is visible from the cell where the observer is located.

The problem of deciding if two cells are visible from each other can be formulated as a linear programming problem. Let us analyze, as an example, the situation in Figure 5.18. Cell 5 is connected through the

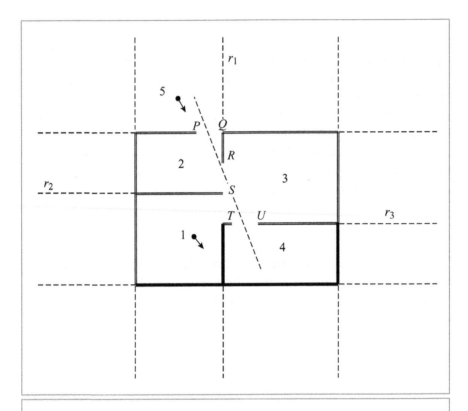

Figure 5.18: Visualization of building enviroments.

door PQ to cell 2, which is connected to cell 3 through the door RS, which, in turn, is connected to cell 4 through the door TU. Cells 4 and 5 are visible from each other when there exists a line, of equation $y = ax + b$, separating points P and Q, points R and S, and points T and U. That is, there must exist real numbers a and b such that the following inequalities are satisfied:

$$\begin{cases} y_P \geq ax_P + b \\ y_Q \leq ax_Q + b \\ y_S \geq ax_S + b \\ y_R \leq ax_R + b \\ y_T \geq ax_T + b \\ y_U \leq ax_U + b \end{cases}.$$

The problem of verifying if a set of linear inequalities has a solution can be solved using a linear program, as studied in Section 5.5.

To construct the visibility graph, it is not necessary to test all the pairs of cells. We can use a propagation strategy. That is, first we identify visible cells that are adjacent, followed by the pairs of cells that are visible through a single cell, and so forth.

For huge scenes, the use of this strategy allows us to reduce significantly the visualization time. Teller and Sequin (1991) report reductions of up to 94% in the number of polygons to be visualized.

5.7 COMMENTS AND REFERENCES

A very intuitive introduction to the conjugate gradient method is the technical report by Shewchuk (1994). A tutorial on least square methods with applications to computer graphics is given in the SIGGRAPH course notes by Pighin and Lewis (2007).

Continuous optimization is an essential ingredient in computer vision, especially for procedures that involve parameter estimation, such as camera calibration or bundle adjustment (i.e., the simultaneous calibration of several cameras that observe the same scene). The book by Hartley and Zisserman (2004) describes several of such applications.

In computer graphics, the use of continuous optimization is more recent. One of the earliest trendsetting papers to propose the use of optimization techniques is that of Gleicher and Witkin (1992), where the authors use constrained optimization to control the movement of a camera.

The area of computer animation makes heavy use of continuous optimization techniques. The main problems fall into two broad categories: 1) the estimation of time-varying parameters that minimize some energy functional and 2) the computation of optimal motions that are subject to constraints. The first problem is the case of unconstrained continuous optimization and can be either linear or nonlinear. The second problem is the case of a continuous optimization with restrictions.

A classical paper related to optimization with restrictions is "Spacetime Constraints," where the problem is formulated as the simulation of a physically valid motion that interpolate key camera poses specified by the animator (Witkin and Kass, 1988).

Another important problem, that concerns character animation is called *rigging*. This technique allows the control of a character through a set of handles. Usually, these handles are part of the character skeleton. Rigging is associated with the *skinning* and *enveloping* techniques. Basically, all these techniques build an association between motion control handles and the geometry of a character. In this way, the shape of the character can be animated and deformed through the influence of the handles. The problem is that constructing a good set of handles is a nontrivial task. Moreover, once these handles are defined, their action on the geometry of the character is also nontrivial. Both aspects of this problem can be solved using continuous optimization methods. Some recent papers in the area are those of Baran and Popovic (2007), Wang *et al.* (2007), Shi *et al.* (2007), Igarashi *et al.* (2005), Sumner *et al.* (2007), and Au *et al.* (2007).

BIBLIOGRAPHY

Au, O. K.-C., H. Fu, C.-L. Tai, and D. Cohen-Or. Handle-aware isolines for scalable shape editing. *ACM Transactions on Graphics*, 26(3):83, 2007.

Baran, I., and J. Popovic. Automatic rigging and animation of 3D characters. *ACM Transactions on Graphics*, 26(3):72, 2007.

Carvalho, P. C. P., F. Szenberg, and M. Gattass. Image-based modeling using a two-step camera calibration method. *IEEE Transactions on Robotics and Automation*, 3(4):388–395, 1987.

Chvatal, V. *Linear Programming.* W. H. Freeman. New york, NY, USA, 1983.

Faugeras, O. *Three-Dimensional Computer Vision: A Geometric Viewpoint.* Cambridge, MA: MIT Press, 1993.

Fuchs, H., Z. M. Kedem, and B. Naylor. On visible surface generation by a priori tree structures. *Computer Graphics (SIGGRAPH '80 Proceedings)*, 14(4):124–133, 1980.

Gleicher, M., and A. Witkin. Through-the-lens camera control. *Computer Graphics*, 26(2):331–340, 1992.

Golub, G. H., and C. F. V. Loan. *Matrix Computations.* Baltimore, MD: Johns Hopkins, 1983.

Hartley, R., and A. Zisserman. *Multiple View Geometry in Computer Vision.* Second edn. New York, NY: Cambridge University Press, 2004.

Igarashi, T., T. Moscovich, and J. F. Hughes. As-rigid-as-possible shape manipulation. *ACM Transactions on Graphics*, 24(3):1134–1141, 2005.

Karmarkar, N. A new polynomial time algorithm for linear programming. *Combinatorica*, 4:373–395, 1984.

Khachian, L. G. A polynomial algorithm for linear programming. *Doklady Akademii Nauk USSR*, 244(5):1093–1096, 1979.

Levenberg, K. A method for the solution of certain non-linear problems in least squares. *Quarterly of Applied Mathematics*, 2:164–168, 1944.

Luenberger, D. G. *Linear and Nonlinear Programming.* Norwell MA: Addison-Wesley, 1984.

Marquardt, D. An algorithm for least-squares estimation of nonlinear parameters. *SIAM Journal of Applied Mathematics*, 11:431–441, 1963.

Matos, A., J. Gomes, A. Parente, H. Siffert, and L. Velho. The visorama system: A functional overview of a new virtual reality environment. In: *Computer Graphics International '97 Proceedings*, 1997.

McMillan, L., and G. Bishop. Plenoptic modeling: An image-based rendering system. *SIGGRAPH '95 Proceedings*, 39–46, 1995.

Meggido, N. Linear programming in linear time when the dimension is fixed. *Journal of ACM*, 31:114–127, 1984.

Pham, B., and Pringle, G. Color correction for an image sequence. *IEEE Computer Graphics and Applications*, 15:38–42, 1995.

Pighin, F., and J. P. Lewis. *Practical Least Squares for Computer Graphics*. SIGGRAPH Course Notes, 2007.

Press, W. H., B. P. Flannery, S. A. Teukolsky, and W. T. Vetterling. *Numerical Recipes in C*. Published in Cambridge University Press, Cambridge, UK, 1988.

Rudin, W. *Principles of Mathematical Analysis*. McGraw-Hill, New York, NY, USA, 1976.

Shewchuk, J. R. *An Introduction to the Conjugate Gradient Method Without the Agonizing Pain*. Technical Report Carnegie Mellon University, 1994(Aug).

Shi, X., K. Zhou, Y. Tong, M. Desbrun, H. Bao, and B. Guo. Mesh puppetry: Cascading optimization of mesh deformation with inverse kinematics. *ACM Transactions on Graphics*, 81:1–10, 2007.

Strang, G. *Linear Algebra and its Applications*. Belmont, CA: Saunders, 1988.

Sumner, R., J. Schmid, and M. Pauly. Embedded deformation for shape manipulation. *ACM Transactions on Graphics*, 26(3):4–50, 2007.

Szeliski, R., and H. Shum. Creating full view panoramic image mosaics and environments. *SIGGRAPH '97 Proceedings*, 251–258, Redmond, WA: 1997.

Teller, S. J., and C. H. Sequin. Visibility preprocessing for interactive walkthroughs. *Computer Graphics (SIGGRAPH '91 Proceedings)*, 25(4):61–69, 1991.

Tsai, R. Y. A versatile camera calibration technique for high accuracy 3D machine vision metrology using off-the-shelf cameras and lenses. *IEEE Transactions on Robotics and Automation*, 3(4), 1987.

Wang, R. Y., K. Pulli, and J. Popovic. Real-time enveloping with rotational regression. *ACM Transactions on Graphics*, 73:1–9, 2007.

Watt, A. *3D Computer Graphics*. Boston, MA: Addison-Wesley, 2000.

Witkin, A., and M. Kass. Spacetime constraints. *Pages 159–168 of SIG-GRAPH '88: Proceedings of the 15th Annual Conference on Computer Graphics and Interactive Techniques*. ACM Press, New York, NY: 1988.

6 COMBINATORIAL OPTIMIZATION

In this chapter, we study optimization problems in the discrete domain where the elements are defined by combinatorial relations. As we will see, besides the problems that are intrinsically discrete, several continuous problems can be discretized and solved in an efficient way using the methods studied here.

6.1 INTRODUCTION

As described in the overview of optimization problems, Chapter 2, in combinatorial optimization, the solution set S is finite. Therefore, we have two possibilities to specify S. The first is a direct specification, using a simple enumeration of its elements. The second consists of an indirect specification, using relations that exploit the problem structure.

The first option, enumerating all the elements of S, besides being clearly uninteresting, may be impractical when the solution set has many elements. Therefore, we are interested in optimization problems where the set S is finite and can be characterized by the use of combinatorial relations based on structural properties of the solution set S. This allows us to obtain a compact description and devise efficient strategies to compute the solution.

6.1.1 SOLUTION METHODS IN COMBINATORIAL OPTIMIZATION

There are two classes of methods to solve a combinatorial optimization problem: specific methods and general methods. We are interested here in the general methods because of their wide range of applicability. Among the general methods, we can establish the following classification:

- Exact methods:
  - Dynamic programming
  - Integer programming
- Approximated methods:
  - Graph cuts
  - Heuristic algorithms
  - Probabilistic algorithms.

For exact methods, in general, it is possible to exploit the combinatorial structure to obtain exact solutions using one of the strategies listed above. Such methods often result in algorithms of polynomial complexity that guarantee the computation of the global minimum for the problem.

Nevertheless, there are certain types of problems for which no efficient algorithm to compute an exact solution is known. In principle, the only option to find the global minimum would be the exhaustive investigation of all the possibilities, which, as mentioned above, is not feasible in practice. When faced with this difficulty, we can use approximate methods that provide a good solution but cannot guarantee that the solution is optimal. In many situations, this result is acceptable. Among the approximate methods, there is a class called *strong approximate methods*. Such methods provide a local solution that is guaranteed to stay close to the global optimal solution, within some bounds.

In this chapter, we study exact methods for combinatorial optimization; in particular, we cover the methods of dynamic and integer programming. We also study graph cuts, which belongs to the class of strong approximate methods. In Chapter 7, we study generic approximate methods to solve both discrete and continuous problems.

We present now a concrete example of a combinatorial problem.

Example 20. Consider Figure 6.1 that represents a network with possible paths connecting the point A to the point B. In this network, as indicated in the figure, each arc is associated with the cost of traversing the path from the initial point to the final point of the arc.

The combinatorial optimization problem, in this example, consists in finding the minimal cost path starting from A and arriving at B.

Two relevant questions to consider are the following:

- What is the size of the problem description?
- What is the size of the solution set S?

We answer the first question using the fact that a network with $(n + 1) \times (n + 1)$ points has $2n(n + 1)$ arcs. Since the network structure is fixed for this class of problems, we only need to specify the costs associated with each arc. That is, the size of the description has an order n^2, which is denoted by $O(n^2)$.

To answer the second question, we observe that in order to go from A to B, it is necessary to traverse, from left to right, $2n$ arcs. In the trajectory, at each node, we have two options: move upward or downward. Moreover, from the network structure, the valid paths must have the same number of upward and downward motion (which is n). That is,

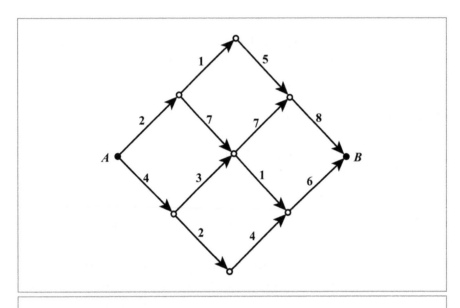

Figure 6.1: Network of connections between *A* and *B*.

the paths have $2n$ arcs, with n arcs moving upward and n moving downward, which results in $\binom{2n}{n}$ possible solutions. Clearly, the size of the solution set S is much larger than n^2.

The previous analysis shows that the size of the combinatorial description of the problem is much smaller than the size of the solution set.

6.1.2 COMPUTATIONAL COMPLEXITY

The computational complexity of an algorithm can be defined as the number of operations necessary to solve a problem whose description has size n. This number is provided as a function of n.

Consider two functions $f\colon \mathbb{N} \to \mathbb{R}$ and $g\colon \mathbb{N} \to \mathbb{R}$. We say that f is $O(g)$, when there exist numbers $K > 0$ and $N \in \mathbb{N}$ such that $f(n) \leq Kg(n)$, for every $n \geq N$.

The function $f = 3n^3 + 2n^2 + 27$, for example, is $O(n^3)$.

An important class is the so-called polynomial algorithms, which have complexity $O(n^p)$, for p fixed.

We are interested in solving efficiently the optimization problem. That is, we need to find algorithms for which the number of operations is comparable to the combinatorial size of the problem. More specifically, we must try to find algorithms of polynomial complexity with respect to the size of the input data.

6.2 DESCRIPTION OF COMBINATORIAL PROBLEMS

One of the most important features of combinatorial optimization methods is format description of the problem data. As we have mentioned previously, the adequate structuring of the data will allow us to obtain an efficient solution to the problem.

Graphs are the appropriate mathematical objects to describe the relations among the data of a combinatorial problem. The network structure used in Example 20 is a specific case of a graph. The combinatorial optimization methods for the different classes of problems will use graphs of different types and exploit their properties in the search for an efficient solution to the problem. For this reason, we provide, in next section, a brief review of the main concepts of graph theory.

6.2.1 GRAPHS

A *graph* is a pair $G = (V, A)$, where $V = \{v_1, \ldots, v_n\}$ is the set of *vertices* and $A = \{v_i, v_j\}, v_i, v_j \in V$ is a set of nonordered pairs of vertices, called *edges* of the graph. When two vertices are connected by an edge, they are called *adjacent* vertices.

It is possible to represent a graph geometrically by representing each vertex $v_j \in V$ as a point in some Euclidean space and each edge $\{v_i, v_j\}$ by a continuous path of the space connecting the vertices v_i and v_j.

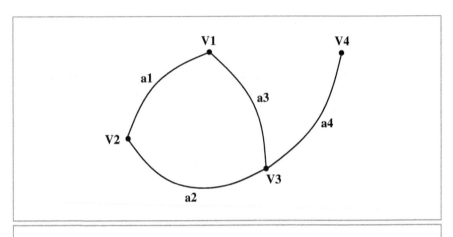

Figure 6.2: Graph.

Figure 6.2 shows a graph with four arcs and four vertices.

A graph is said to be *oriented* when the pairs of vertices are ordered pairs. In this case, the edges are called *arcs* of the graph.

Figure 6.3 shows an oriented graph with five arcs and four vertices.

Now we describe some structures used to represent a graph.

6.2.2 ADJACENCY MATRIX

A graph can be represented by a matrix that describes the adjacencies between the vertices of the graph. That is, for each vertex v_j, it provides the other vertices that are connected to v_j by an edge. Note that we have to establish an ordering for the graph vertices in order to associate them with the indices of the matrix.

Therefore, we have a matrix $M_{|V| \times |V|} = (a_{ij})$, where

$$a_{ij} = \begin{cases} 1 & \text{if } ij \in A \\ 0 & \text{if } ij \notin A \end{cases}.$$

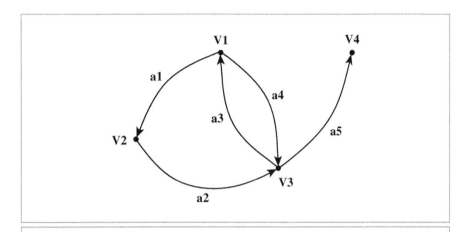

Figure 6.3: Oriented graph.

Note that all the elements on the diagonal of the adjacency matrix are equal to zero. Moreover, the adjacency matrix of a nonoriented graph is always symmetric, and the adjacency matrix of an oriented graph is not symmetric, in general.

Below, we show the adjacency matrix of the graph in Figure 6.2 (left), and the adjacency matrix of the graph in Figure 6.3 (right).

$$\begin{pmatrix} 0 & 1 & 1 & 0 \\ 1 & 0 & 1 & 0 \\ 1 & 1 & 0 & 1 \\ 0 & 0 & 1 & 0 \end{pmatrix}; \qquad \begin{pmatrix} 0 & 1 & 1 & 0 \\ 0 & 0 & 1 & 0 \\ 1 & 0 & 0 & 1 \\ 0 & 0 & 0 & 0 \end{pmatrix}.$$

6.2.3 INCIDENCE MATRIX

Another useful graph representation is the so-called incidence matrix. This matrix provides, for each vertex v_j, the edges $\{v_j, v_k\}$, $v_k \in A$, i.e., the edges that are incident to vertex v_j (the edges that connect v_j to some other vertex).

Therefore, we have a matrix $K_{|V| \times |A|} = (a_{ij})$, where

$$a_{ij} = \begin{cases} 1 & \text{if the edge } j \text{ is incident to the vertex } i \\ 0 & \text{otherwise} \end{cases}.$$

Below, we show the incidence matrix of the graph in Figure 6.2:

$$\begin{pmatrix} 1 & 0 & 1 & 0 \\ 1 & 1 & 0 & 0 \\ 0 & 1 & 1 & 1 \\ 0 & 0 & 0 & 1 \end{pmatrix}.$$

For an oriented graph, the incidence matrix must indicate also the arc direction. In this case, the incidence matrix is $K_{|V| \times |A|} = (a_{ij})$, where

$$a_{ij} = \begin{cases} 1 & \text{if edge } j \text{ starts at vertex } i \\ -1 & \text{if edge } j \text{ ends at vertex } i \\ 0 & \text{if edge } j \text{ is not incident to vertex } i \end{cases}$$

Below, we show the incidence matrix of the graph in Figure 6.3:

$$\begin{pmatrix} 1 & 0 & -1 & 1 & 0 \\ -1 & 1 & 0 & 0 & 0 \\ 0 & -1 & 1 & -1 & 1 \\ 0 & 0 & 0 & 0 & -1 \end{pmatrix}.$$

6.2.4 NETWORKS

In many problems of combinatorial optimization, we use a special kind of graph called flow network.

A *flow network* $G(V, A)$ is defined as a fully connected directed graph where each edge $(u, v) \in A$ has a positive weight $c(u, v) \geq 0$, also known as *edge capacity*.

The network has two special vertices designated the source s and the sink t.

A flow in the network G is a real-valued function $f : V \times V \to \mathbb{R}$, such that the following properties are satisfied:

- Capacity constraint:
  for all $u, v \in V, f(u, v) \leq c(u, v)$.
- Skew symmetry:
  for all $u, v \in V, f(u, v) = -f(v, u)$.
- Flow conservation:
  for all $u, v \in (V - \{s, t\})$, $\sum_{v \in V} f(u, v) = 0$.

The value of the flow is defined as $|f| = \sum_{v \in V} f(s, v)$. In other words, it is the total flow out of the source node in the flow network G.

6.3 DYNAMIC PROGRAMMING

The method of dynamic programming, also called recursive or sequenced optimization, applies to combinatorial problems that can be decomposed into a sequence of stages. The basic operation in each of these stages consists in taking a partial decision (local decision), which will lead us to the minimum solution to the problem.

Dynamic programming is based on the *optimality principle*, which is stated below.

Proposition 2 [optimality principle]. *The best sequence of decisions has the property that whatever the last decision we have taken, the sequence of decisions that led us to the next to the last stage must also be optimal.*

The optimality principle allows us to elaborate a sequential algorithm whose solution strategy at each stage depends only on a *local* decision. That is, by supposing that we have a partial optimal sequence of decisions, the current stage depends only on a localized decision, i.e., a decision that is only based on part of the available information.

In order to describe the method in more detail, we introduce the appropriate formalism and methodology.

By assuming that the problem can be decomposed into a sequence of stages, the method will use

- *stage variables* and
- *state variables*.

We should remark that each stage can be associated with the time or the evolution of the algorithm toward the solution.

Therefore, we may suppose that the problem has N stages, and each stage has a set X_n of states that describe the partial result obtained.

Then, we define the function $f_n(x)$ that provides the optimal value to attain at the state x of stage n.

Using the optimality principle, we obtain the *fundamental equation* of dynamic programming:

$$f_{n+1}(x) = \min_{u \in \text{etapa } n} \{f_n(x) + c(u, x)\},$$

where $c(u, x)$ is the transition cost from state u in stage n to state x in stage $n + 1$.

The Equation 6.3 provides the basic mechanism of recursion of the method and at the same time indicates the computation to be used in the algorithm sequence. More precisely, at each stage n varying from 1 to N, the algorithm computes the function $f_n(x)$ for every state variables $x \in X_n$ and registers the transitions $u \to x$ associated with them. At the end of this process, we have computed the cost of the optimal path. Thus, we can traverse the reverse trajectory, using the inverse of the transitions $u \to x$, and obtain the shortest path. This last operation is called *backtrack*.

The algorithm needs a number of operations proportional to the number N of stages multiplied by the number of transitions. By supposing that we have M possible transitions at each stage, the complexity of the algorithm is $O(NM)$.

In order to illustrate the algorithm, we give a concrete example below.

Example 21. Consider the diagram in Figure 6.4, which shows a network of connections between A and B with the associated costs at each arc. The optimization problem consists in computing the minimal cost path between A and B.

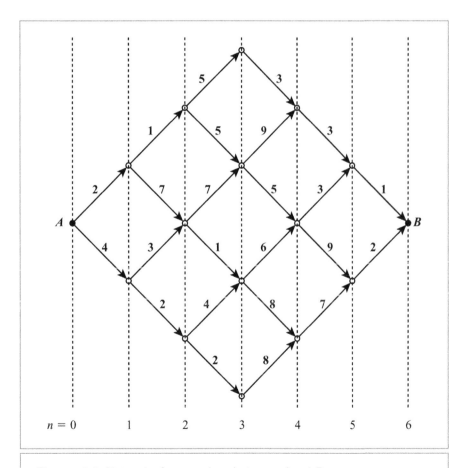

Figure 6.4: Network of connections between A and B.

From the combinatorial structure of the network, it is possible to decompose the problem into a sequence of stages that correspond to the partial paths to traverse the network from left to right. Note that these paths start at A and end in the intermediate vertices aligned with the vertical lines. In Figure 6.4, these lines are represented by dashed straight lines.

At each stage of the algorithm, we compute the cost of all optimal partial paths up to the current stage, along with the pointers indicating the optimal transition.

In Figure 6.5, we add the value of the minimum cost $f_n(x)$ associated with each vertex x of the network, as well as a pointer indicating

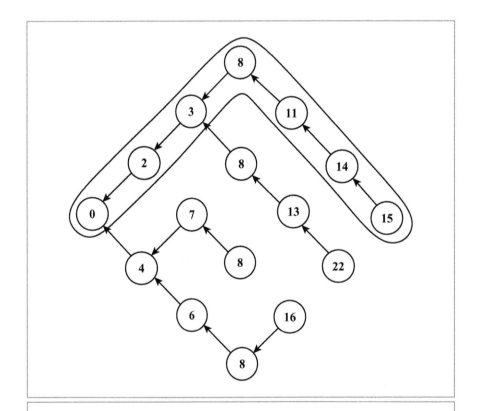

Figure 6.5: State variable of the problem.

the vertex u in the previous stage associated with the minimum path to that vertex. Moreover, we draw a thin curve around the shortest path.

6.3.1 CONSTRUCTING THE STATE OF THE PROBLEM

According to the results from previous section, the method of dynamic programming for combinatorial optimization problems is structured based on two types of information: stage variables and state variables.

The solution to the problem develops sequentially (by stages), using the recursion originated from the optimality principle. At each stage, we have to maintain information about previous stages (the state of the problem) in order to advance and compute the solution in the next stage.

An important consequence of the optimality principle is that we need only local information in our computations. The nature and the scope of the state information depend on the problem to be solved. In fact, this is one of the most important aspects of modeling in dynamic programming.

In order to better illustrate the formulation and the use of the state variables, we give another example.

Example 22. Consider Example 21 in which we need to compute the minimum cost of path in a network connecting the points A and B, with costs associated with each arc. We now modify the problem by including one more component in the cost of a path: each change in direction will imply a penalty (an additional cost of c units).

In Example 21, the decision rule is given by

$$f(n,y) = \min \begin{cases} f(n-1,y-1) + u(n-1,y-1) \\ f(n-1,y+1) + d(n-1,y+1), \end{cases}$$

where $u(n, y)$ and $d(n, y)$ are the costs of the arcs connecting, respectively, $y - 1$ to y (in the direction "up") and $y + 1$ to y (in the direction "down"). The only state information stored at each node (n, y) of the network is the optimal cost $f(n, y)$ of the partial path from A to (n, y).

In order to consider in the cost the changes of direction of the path, we need more state information. Now, the state is given by $f(n, y, a)$, which is the value of the best path to arrive at the node (n, y) from the direction a. In order to arrive at node (n, y), it is possible to come either from $(n - 1, y - 1)$ or from $(n - 1, y + 1)$; we denote these directions, respectively, by $\nearrow$ and $\searrow$. Thus, at each node, we have to store the value of $f(n, y, a)$ for these two directions.

After including in the cost computation the change of direction, we obtain

$$f(n, y, \nearrow) = \min \begin{cases} f(n - 1, y - 1, \searrow) + c + u(n - 1, y - 1) \\ f(n - 1, y - 1, \nearrow) + u(n - 1, y - 1) \end{cases}$$

and

$$f(n, y, \searrow) = \min \begin{cases} f(n - 1, y + 1, \searrow) + d(n - 1, y + 1) \\ f(n - 1, y + 1, \nearrow) + c + d(n - 1, y + 1). \end{cases}$$

We should remark that since in the computation of stage n we are using the value of f in stage $n - 1$, we consider the change of direction between the stages $n - 2, n - 1$, and $n - 1, n$.

6.3.2 REMARKS ABOUT GENERAL PROBLEMS

The adequate modeling of a combinatorial optimization problem relative to the method of dynamic programming requires that we are able to express its solution by a sequence of stages where, in each stage, we use only the localized state information.

In the examples given, we defined the recursion from the end to the beginning. That is, we computed the state of stage n based on stage

$n + 1$. We can think about the solution to this problem also from the beginning to the end. We leave this as an exercise to the reader.

Another remark is that the solution to the problem corresponds essentially to filling a table where the columns correspond to the stages $n = 1, \ldots, N + 1$, and the rows to the states $x = 0, \ldots, M + 1$. One of the consequences of this fact is that it is possible to implement the method using an electronic spreadsheet program.

6.3.3 PROBLEMS OF RESOURCE ALLOCATION

One class of practical problems that can be solved using dynamic programming is the class of resource allocation problems.

Knapsack Problem

The *knapsack problem* is the simplest version of the general problem of resource allocation. Intuitively, the knapsack problem consists in choosing among a set of possible objects that can fit into one bag to be carried on a trip, such as to maximize their aggregate benefit, considering that each object has a weight and the bag has a limited capacity.

The precise formulation of the problem is given by

$$\begin{aligned} \text{maximize} \quad & c_1 x_1 + \cdots + c_n x_n \\ \text{subject to} \quad & w_1 x_1 + \cdots + w_n x_n \leq L. \end{aligned}$$

There are two versions for the problem:

- zero-one version: $x_i \in \{0, 1\}$
- linear version: $x_i \geq 0, x_i \in \mathbb{Z}$.

The recursion equation is given by

$$s(k, y) = \max \begin{cases} s(k - 1, y - w_k) + c_k, \\ s(k - 1, y) \end{cases}$$

where $s(k, y)$ is the knapsack of maximum value with size smaller than or equal to y, using only the first k objects. In the first case, we put the element k in the knapsack. In the second case, we keep the previous knapsack, i.e., without the element k.

The objective of the problem consists in finding $s(n, L)$, given the initial conditions

$$s(0, y) = 0, \quad \forall y$$
$$s(k, y) = \infty, \quad \text{if } y < 0.$$

The complexity of the problem is $O(nL)$. At first sight, it would appear that this algorithm runs in polynomial time. However, storing an integer M requires only $\log n$ bits. Thus, the encoding size of a knapsack problem is $O(n \log L)$. Hence, the running time of the algorithm above is not polynomial (it is called a *pseudopolynomial* algorithm).

There is a decision problem that can be associated with the knapsack problem and is stated as follows: "Given items of different weights and a knapsack, is there a subset that exceeds a certain value?" This decision problem is *NP*-complete.

6.4 SHORTEST PATHS IN GRAPHS

In order to generalize the method of dynamic programming, we study shortest path problems in general graphs (oriented or not).

6.4.1 SHORTEST PATHS IN ACYCLIC ORIENTED GRAPHS

Acyclic oriented graphs, or directed acyclic graphs (DAGs), constitute one of the richest structures to describe combinatorial problems.

Theorem 15 [ordering of DAGs]. *The nodes in an acyclic oriented graph can be numbered from 1 to n, in such a way that every arc is of the form ij with i < j.*

The proof is constructive, and we leave it as an exercise to the reader.

Given an acyclic oriented graph $g = (V, E)$, we want to find a shortest path from s to t, where $s, t \in V$. We may suppose, without loss of generality, that $s = 1$ and $t = |V|$.

Algorithm 1

---

1. Order the nodes in such a way that the arc $ij \Rightarrow i < j$,
   or verify that the graph is not acyclic and leave the algorithm.
2. Let c_i the length of the smaller path from 1 to i.
   Then: $c_1 = 0$

for $i = 2, \ldots, n$ **do**
   $c_i = \min_{j < i} \{c_j + a_{ji}\}$
end for

We denote by a_{ji} the cost of the arc that goes from j to i. Note that if there does not exist an arc from j to i in the graph, we can take $a_{ji} = \infty$.

In order to study the complexity of the algorithm, consider that the graph has n arcs. In the worst case, all the arcs are connected to each other. Thus, the complexity is $O(|V|^2)$ or $O(|E|)$ depending on the representation of the graph.

Note that the second case requires sparse storage. For sparse graphs, it is common to replace the incidence matrix with sparse structures, such as adjacency lists.

Remark. The recursion order is established from the incidence relations, and the method is equally applicable to shortest or longest paths, independently of the cost signals.

6.4.2 DIRECTED GRAPHS POSSIBLY WITH CYCLES

In this section, we study problems related to shortest paths in directed graphs possibly with cycles. In order that the problem has an optimal solution, we must assume the following hypothesis. *There are no cycles with negative (positive) total cost for problems of shortest (longest) path.*

In the presence of negative cycles, the shortest path problem is NP-complete (it may even be unbounded if any path from the source to the destination reaches such a cycle).

Some classes of problems with negative cycles can be solved using intelligent enumeration methods such as *branch and bound*. These methods, although do not have polynomial complexity, in practice are reasonably efficient from the computational point of view.

Now we concentrate in the case without cycles, and we postpone the discussion of the case with cycle to the next section.

The optimality principle for the shortest path in graphs without negative cycles can be formulated in a concise way using Bellman equation:

$$c_j = \min_{i \neq j} \{c_i + a_{ij}\}, \tag{6.1}$$

where c_j is the length of the smallest path from 1 to j.

Remark. A shortest path from s to t has at most $|V| - 1$ arcs. That is, the path has at most $|V| - 2$ intermediate nodes.

We describe in what follows a shortest-path algorithm for directed graphs that satisfy the hypothesis of not having cycles with total negative cost.

Definition of the Problem

Let $c_k(j)$ denote the length of the shortest path from 1 to j with, at most, k arcs, and a_{ij} be the cost of transition from i to j. If there exists an arc ij, this cost is given by the problem, otherwise the cost is ∞.

Algorithm 2
---

1. **Initialization:**
   $$c_1(1) = 0$$
   $$c_1(j) = a_{1j} \text{ for } j \neq 1$$
2. **Recursion:**
   $$c_k(j) = \min_{i \neq j}\{c_{k-1}(i) + a_{ij}\}$$

We want to obtain $c_{|V|-1}(j)$.

The complexity of the problem is $O(|V|^3)$ or $O(|V||A|)$.

Theorem 16. *A graph has no negative cycles if and only if* $c_{|V|}(j) = c_{|V|-1}(j)$ *for every j.*

Proof. Suppose that there are no negative cycles. A path with $|V|$ arcs has repeated nodes and therefore has length greater than or equal to a subpath with less than $|V|$ arcs. Thus, $c_{|V|}(j) = c_{|V|-1}(j)$ for every j.   □

Algorithm 2 also computes, with slightly greater complexity, the shortest paths from i to j for every pair ij.

$$c_k(i, j) = \min_l\{a_{il} + c_{k-1}(lj)\}.$$

The complexity is $O(|V|^3 \log |V|)$ or $O(|V||A| \log |V|)$. This problem is known as the *multiterminal* problem.

6.4.3 DIJKSTRA ALGORITHM

This algorithm assumes that all costs are nonnegative.

Recalling Bellman equations:

$$c(1) = 0$$
$$c(j) = \min_{i \neq j}\{c(i) + a_{ij}\},$$

where c is a function that attributes a value for each node of the graph.

It is not difficult to prove that c satisfies Bellman equations if and only if $c(j)$ is the length of the shortest path from 1 to j.

Dijkstra algorithm computes $c(i)$ in an order $v_1, v_2, \ldots v_{|V|}$, such that $c(v_1) \leq c(v_2) \leq \ldots \leq c(v_{|V|})$

Algorithm 3 Dijkstra

---

1. **Initialization:**
   $c_1(1) = 0$ and $c_1(j) = a_{1j}$ for $j \neq 1$
   $p = \{1\}, \overline{p} = \{2, \ldots, |V|\}$
2. **Recursion:**
   while $\overline{p} \neq \varnothing$ **do**
     Choose k such that $c(k) = \min_{j \in \overline{p}} c(j)$
     $p = p \cup \{k\}$
     $\overline{p} = \overline{p} - \{k\}$
     for all $j \in \overline{p}$ **do**
       $c(j) = \min\{c_j(i), c(k) + a_{kj}\}$
     end for
   end while

Theorem 17. *At the end of each iteration, $c(j)$ is the length of the shortest path from 1 to j using only intermediate nodes in p. Thus, when all vertices have been included in p, $c(j)$ is the length of the shortest path from 1 to j.*

The proof is by induction.

The complexity of the problem is $O(|V|^2)$ or $O(|A| \log |V|)$.

6.5 INTEGER PROGRAMMING

In this section, we study problems on integer programming, which appear when we restrict linear continuous optimization problems to the set of integer numbers.

6.5.1 LINEAR AND INTEGER PROGRAMMING

In Chapter 5 on continuous optimization, we defined a linear programming problem as follows:

$$\min \sum_{j=1}^{n} c_j x_j$$

$$\text{subject to } \sum_{j=1}^{n} a_{ij} x_j \leq b_i,$$

for $i = 1, \ldots m$.

This is a problem with n variables and m constraints, in which both the objective function and the constraints themselves are linear functions.

We remark that other inequality or equality constraint can be added to the above equations. Moreover, the variables have no restrictions on their values or they might have constraints such as $x_j \leq 0$; $x_j \geq 0$ or $x_j \in [l_j, u_j]$, which naturally configures itself as constraints of the problem.

An *integer program* is a linear program with the additional constraint $x_j \in \mathbb{Z}$.

In what follows, we show that an integer program can be naturally formulated as a combinatorial optimization problem and vice versa.

Example 23 [shortest path]. Consider the graph described by its incidence matrix

$$\begin{pmatrix} 1 & 1 & 0 & 0 & 0 & 0 \\ -1 & 0 & -1 & 0 & 0 & 0 \\ 0 & -1 & 1 & -1 & 0 & 1 \\ 0 & 0 & 0 & 0 & -1 & -1 \end{pmatrix},$$

where the values 1 and -1 at the entry ij of the matrix indicate, respectively, that the arc j arrives or leaves the node i.

We want to find the shortest path between the initial node s and the final node t. In the formulation of the problem as an integer program, we have

- variables:

$$x_j = \begin{cases} 1, & \text{if the arc } j \text{ is part of the path} \\ 0, & \text{otherwise} \end{cases}$$

- objective function:

$$\min \sum c_j x_j$$

- constraints:

$$\sum a_{ij} x_j = \begin{cases} 1, & \text{if } i = s \\ -1, & \text{if } i = t \\ 1, & \text{if } i \neq s, t. \end{cases}$$

6.5.2 MINIMUM COST FLOW PROBLEM

This is a more general case of the shortest path problem.

For each node, we have d_i, which is the available amount ($d_i > 0$) or the demanded amount ($d_i < 0$) associated with the node i.

For each arc, we have μ_j, which is the *capacity* of the arc.

The minimum cost flow problem consists in

$$\min \sum c_j x_j$$

$$\text{subject to } \sum a_{ij} x_j = d_i$$

with $0 \leq x_j \leq \mu_j$.

Intuitively, it consists in finding the cheapest way of sending a certain amount of flow through a network.

A variation of this problem is to find a flow that is maximum but has the lowest cost among the maximums. This is called a *minimum-cost maximum-flow problem*.

6.5.3 LINEAR PROGRAMS WITH INTEGER SOLUTIONS

Theorem 18. *A flow problem of minimum cost with integer demands and capacities has an optimal integer solution.*

$$\min c_j x_j$$

$$\text{subject to } \sum_j a_{ij} x_j = d_i, \quad i = 1, \ldots |V|$$

$$\text{for } 0 \geq x_j \geq \mu_j,$$

where a_{ij} are the elements of the incidence matrix A, and x_j corresponds to the amount sent along the arc j.

Proof.   We will prove that noninteger solutions are not basic.

Let $\bar{x}$ be a noninteger solution to the problem. There exists a circle composed of arcs where $\bar{x}_j$ is a fractionary.

Let $\bar{x}(\varepsilon)$ be a new solution given by

$$\bar{x}_j(\varepsilon) = \begin{cases} \bar{x}_j, & \text{if } j \notin \text{cycle} \\ \bar{x}_j + \varepsilon, & \text{if } j \in \text{cycle with equal orientation} \\ \bar{x}_j - \varepsilon, & \text{if } j \in \text{cycle with opposite orientation.} \end{cases}$$

$\bar{x}(\varepsilon)$ satisfies the conservation equations.

$\bar{x}(\varepsilon)$ satisfies $0 \geq \bar{x}(\varepsilon)$ of μ_j for $|\varepsilon|$ sufficiently small.

Since $\bar{x} = \frac{1}{2}\bar{x}(\varepsilon) = \frac{1}{2}\bar{x}(-\varepsilon)$, $\bar{x}$ is not a basic solution.   □

Specific cases

Shortest path: one node with $d = 1$, one node with $d = -1$, and $\mu_j = 1$, $\forall j$; bipartite matching (or assignment matching).

Remark. The linear program has an integer solution in this case but not in the case of nonbipartite matching.

6.6 GRAPH CUTS

A *cut* in a graph is a means to partition its vertices into two sets. Formally, let $G(V, A)$ be a graph. A cut $C = (S, T)$ induces a partition of G into the disjoint sets S and T. Consequently, any edge $(v_i, v_j) \in A$, with $v_i \in S$ and $v_j \in T$ (or $v_j \in S$ and $v_i \in T$, in the case of a directed graph), crosses the cut and is called a *cut edge*.

The *size of a cut* is the total number of cut edges. In a weighted graph, the size of the cut is defined to be the sum of weights $w(u, v)$ of the edges crossing the cut and can be interpreted as the cost of making the cut, denoted by $|C|$:

$$\text{cost}(S, T) = |C| = \sum_{u \in S, v \in T, (u,v) \in A} w(u, v). \tag{6.2}$$

Graph cut problems are important for various reasons. First, there is a nontrivial relation between minimum-cut and maximum-flow problems that sheds light on a class of combinatorial optimization algorithms. Second, cuts in graphs provide an intuitive way to formulate problems in many areas, particularly in graphics and vision. Third, it is possible to exploit this problem structure in connection with energy minimization, which gives a powerful way to discretize some continuous optimization problems. Finally, recent developments introduced new algorithms that have many applications in vision.

In this section, we discuss the above issues in more detail.

6.6.1 EQUIVALENCE BETWEEN MIN-CUT AND MAX-FLOW

A natural way to define a cut, $C = (S, T)$, in a network (e.g., a directed graph) is to associate one of the sets with the *source s* and the other set with the *sink t*. In this context, the cut (S, T) with $s \in S$ and $t \in T$ separates the source from the sink. Therefore, a cut in a network is a set of arcs, such that if they are removed, there is no path going from s to t.

Also, in the case of flow problems, the size of a cut is defined as the sum of the capacities of the cut edges in the graph.

One of the fundamental results in combinatorial optimization is the theorem below, which states that the minimum cut and the maximum flow are equivalent problems.

Theorem 19 [max-flow min-cut]. *In every network, the maximum flow equals the minimum capacity of a cut.*

This theorem was proved in 1956 independently by Ford and Fulkerson (1962)) and also by Elias, Feinstein, and Shannon Elias *et al.* (1956). The proof by Ford and Fulkerson is constructive and leads to their algorithm for solving the problem, which we present below. The theorem can also be proved by applying the duality theorem for linear programming (Gale and Tucker, 1951). The reason is that, as we have seen in Section 6.5.2, determining maximum flows is a special kind of linear programming problem.

The intuition behind the max-flow/min-cut theorem is as follows. If we interpret a directed graph as network of water pipes, the maximum flow through this system is the maximum amount of water that can be sent from the source to the sink, considering the flux capacity of the pipes equal to edge weights. The equivalence theorem states that this maximum flow saturates a subset of the edges C in the graph, which divides the nodes into two disjoint parts $\{S, T\}$ such that C corresponds to a minimum cut. In essence, this interpretation says that the "throughput" of a network is determined by its "bottlenecks."

There are two basic strategies to solve the max-flow/min-cut problem:

- Augmenting path methods: these algorithms work by repeatedly finding a nonsaturated path of positive capacity from the source to the sink and adding it to the solution until the maximum flow is reached. The difference between algorithms of this type is *how* they select the augmenting path. The algorithm might have a slow convergence if new paths add little to the total flow. The Ford–Fulkerson algorithm, described below, is the prototypical example of this class of algorithm.
- Preflow-push methods: these algorithms push flow from one node to another, ignoring until the very end that the constraint that the in-flow must be equal to the out-flow at each node of the network. Preflow-push methods can be faster than augmenting path methods because multiple paths can be computed simultaneously. An example of this class of algorithm is the Goldberg–Tarjan method (Goldberg and Tarjan, 1988).

We now take a closer look at the Ford–Fulkerson algorithm. The problem is to find the maximum flow f from the source s to the sink t, given a directed graph $G(V, A)$, with capacity $c(u, v)$ and flow $f(u, v)$ for each edge (u, v).

The algorithm maintains for each step a *legal flow* through the network by enforcing the following restrictions:

- $f(u, v) \leq c(u, v)$, i.e., the flow does not exceed edge capacities
- $f(u, v) = -f(v, u)$, i.e., the net flow is balanced
- $\sum_v f(u, v) = 0 \Leftrightarrow f_{\text{in}}(u) = f_{\text{out}}(u) \forall u \in V - \{s, t\}$.

Information about the current flow at each step of the algorithm is defined through a *residual graph* G_f, which is identical to G except that the capacity $c_f(u, v)$ reflects the residual capacity left for (u, v) given the allocated flow $f(u, v)$ in the edge, i.e., $c_f(u, v) = c(u, v) - f(u, v)$.

Initially, there is no flow, and each edge has its original capacities. At each iteration, the algorithm finds the shortest $s \to t$ path along nonsaturated edges of the residual graph. If a path is found, it is augmented by pushing an amount of flow that saturates at least one of the edges in the path. The maximum flow is reached when any $s \to t$ path crosses one saturated edge in the residual graph. The output of the algorithm is both the maximum flow and the set of saturated edges (i.e., the minimum cut).

Algorithm 4 shows the pseudocode of the method.

The augmenting path can be found by a breadth-first search or a depth-first search in $G_f(V, A)$. When the former is used, the algorithm is called Edmonds–Karp (Edmonds and Karp, 1972) or Dinic (Dinic, 1970).

The complexity of the Ford–Fulkerson algorithm is bounded by $O(Af)$, where A is the number of network edges and f is the maximum flow. The Edmonds–Karp variation of the algorithm has a runtime independent of the maximum flow value with complexity $O(VA^2)$.

Algorithm 4 Ford–Fulkerson

---

for all edges (u, v) initialize $f(u, v) = 0$
while there is a path p from s to t, such that $c_f(u, v) > 0$ for $(u, v) \in p$ **do**
  Find $c_f(p) = \min\{c_f(u, v) | (u, v) \in p\}$
  for each edge $(u, v) \in p$ **do**
    $f(u, v) = f(u, v) + c_f(p)$
    $f(v, u) = f(v, u) - c_f(p)$
  end for
end while

6.6.2 THE LABELING PROBLEM

A cut in a graph can be interpreted as a binary labeling. This is useful in many application areas, such image processing and vision.

More specifically, starting with an original graph describing some application-dependent domain, we can extend it to create an s–t graph

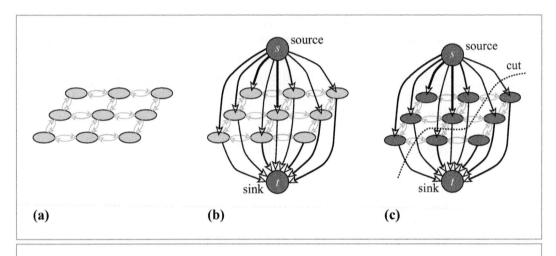

Figure 6.6: Labeling a graph *s–t*.

in order to use graph cuts to label its vertices. Formally, given a graph $G_o(V, A)$, the extended $s–t$ graph is constructed simply by adding two special vertices s and t to the graph and linking all vertices $v \in V$ to both s and t (see Figure 6.6(a)-(b)).

The labeling is represented by a function l mapping from the set of original vertices $V - \{s, t\}$ to $\{0, 1\}$, where $l(v) = 0$ means that $v \in S$ and $l(v) = 1$ means that $v \in T$ (see Figure 6.6(c)).

In this context, a cut is a binary partition of the graph viewed as a binary-valued labeling of its vertices. We remark that while we can solve the binary min-cut problem in polynomial time as seen in Section 6.6.1, the generalization of min-cut to more than two terminals, called multiway cut problem (Dahlhaus *et al.*, 1992), is NP-hard.

6.6.3 REPRESENTING ENERGY FUNCTIONS WITH GRAPHS

It is natural in various applications to represent a problem as the minimization of some energy function. Since given a graph $G(V, A)$, each cut $C = (S, T)$ has some cost, the graph G can be used to encode an energy function in the following way.

Consider the graph $G(V, A)$ with terminals s, t, and $V = \{v_1, \ldots, v_n, s, t\}$. G represents the energy function mapping from all cuts on G to the nonnegative real numbers (i.e., the cost $|C|$). To this end, any cut is described by n binary variables $x_1, \ldots, x_n$, which correspond to vertices in G (excluding s and t). The variable x_i takes a value $x_i = 0$ when $v_i \in S$, and $x_i = 1$ when $v_i \in T$.

Thus, the energy E represented by G is a function $E(x_1, \ldots, x_n)$ of n binary variables, and the value of E is equal to the cost of the cut $|C|$ associated with the configuration $x_1, \ldots, x_n$, where $x_i \in \{0, 1\}$.

A question that arises at this point is: What is the class of energy functions E for which it is possible to construct an $s\!-\!t$ graph G that represents E?

We call this class of energy functions *graph representable* and define them as follows.

Definition 1. *A function E of n binary values is graph representable if there exists an $s\!-\!t$ graph $G(V, A)$ with a subset of vertices $V_0 = \{v_1, \ldots v_n\} \subset V - \{s, t\}$ such that for any configuration $x_1, \ldots, x_n$, the value of $E(x_1, \ldots, x_n)$ is equal to a constant plus the cost of the minimum cut $C(S, T)$ among all cuts in which $v_i \in S \Leftrightarrow x_i = 0$ and $v_i \in T \Leftrightarrow x_i = 1$.*

Kolmogorov and Zabih (2004) give a characterization of the classes of functions F^2 and F^3 that are graph representable and show how to construct the graphs G within these classes. Here we describe the energy functions of class F^2 that has a wide applicability.

The functions of class F^2 are energy functions that can be written as a sum of functions of up to two binary variables.

$$E(x_1, \ldots, x_n) = \sum_i E^i(x_i) + \sum_{i<j} E^{i,j}(x_i, x_j). \tag{6.3}$$

It can be observed that the terms $E^i(x_i)$ measure the cost of assigning a label s or t to the vertex v_i, while the terms $E^{i,j}(x_i, x_j)$ measure the cost of assigning equal or different labels to the pair of vertices v_i, v_j.

Theorem 20 [F^2 theorem]. *Let E be a function of n binary values from the class F^2, as given by Equation 6.3. Then, E is graph representable if and only if each term $E^{i,j}$ satisfies the inequality*

$$E^{i,j}(0, 0) + E^{i,j}(1, 1) \leq E^{i,j}(0, 1) + E^{i,j}(1, 0). \qquad (6.4)$$

The functions satisfying condition 6.4 are called *regular*. Kolmogorov and Zabih (2004) proved that regularity is a necessary and sufficient condition for an energy function if F^2 be graph representable. Their proof is constructive and therefore can be used as a tool[1] to generate the graph G, given an energy function E in F^2.

The strategy behind the proof is to construct each term of the function separately in the form of edges of the graph with appropriate weights and then merge all these subgraphs together to form G. The validity of this construction is justified by the additivity theorem, which states that the sum of two graph-representable functions is also graph representable. We refer the reader to the original paper Kolmogorov and Zabih (2004) for the details.

Energy functions of the form (6.3) arise in various application contexts; in particular, they are used in the Bayesian labeling of first-order Markov random fields (Li, 1995).

One important remark regarding energy functions of the class F^2 is that when the vertices of the graph come from a discretization of the spatial domain of some object, such as an image or a surface, these energy functions can be used to impose spatial smoothness. This is done by penalizing adjacent vertices to have different labels. However,

---

1   In fact, the authors developed a software available on the Web (Kolmogorov, 2004) that takes an energy function as input and automatically constructs the associated graph. Then, it minimizes the energy using graph cuts.

in the cases of interest, the functions are piecewise smooth and have discontinuities on the boundary of regions with different labels. For this reason, it is important that the energy function does not overpenalize such labelings. Functions with this property are called *discontinuity preserving*.

An example of discontinuity-preserving function is given by the Potts model.

$$V(\alpha, \beta) = kT(\alpha \neq \beta), \tag{6.5}$$

where α, β are label assignments and k is a constant, and $T(.) = 1$ if its argument is true and 0 otherwise. This model encourages labelings consisting of several regions where vertices in the same region have equal label.

6.6.4 THE EXPANSION-MOVE ALGORITHM

Up to this point, we have seen that the min-cut method can be used to compute optimal binary labelings. However, for most applications, we are interested in computing n-ary labelings, which is an NP-hard problem, as discussed in Section 6.6.2.

In this subsection, we present an algorithm that can generate n-ary labelings by computing a local minimum in a strong sense of the energy function (Boykov *et al.*, 2001).

This algorithm is called the *expansion-move*, and it minimizes an energy function with nonbinary variables by repeatedly minimizing an energy function with binary variables using min-cut.

The strong local minimum condition guarantees that the local minimum computed by the algorithm lies within a multiplicative factor of the global minimum (this factor is at least 2 and depends only on the binary terms $E^{i,j}$ of the energy function).

The central operation of the expansion-move algorithm consists in finding a minimal $s-t$ cut and is called *α-expansion*. It is defined as follows. Consider a labeling f (partition P) and a particular label α. Another labeling f' (partition P') is an α-expansion move from f if $P_\alpha \subset P'_\alpha$ and $P'_l \subset P_l$ for any label $l \neq \alpha$. In other words, the set of vertices assigned the label α increases when going from f to f'.

The expansion-move algorithm cycles through the labels α in some order (fixed or random) and finds the lowest energy α-expansion move from the current labeling. If this move gives a lower energy than the current labeling, then it replaces the current labeling. The algorithm stops when there is no α-expansion move with lower energy, for any α. The output labeling is a local minimum of the energy with respect to expansion moves.

This algorithm is guaranteed to terminate in a finite number of cycles. In fact, it is possible to prove termination in $O(|P|)$ cycles. However, by the nature of the algorithm, most of the improvement takes place during the first cycle and, in practice, the algorithm terminates after a few cycles.

The pseudocode of the expansion-move algorithm is shown in Algorithm 5.

Algorithm 5 Expansion-Move

Start with an arbitrary labeling f
repeat
    success = false
    for each label $\alpha \in L$ **do**
        Find $\hat{f} = \arg \min E(f')$ among f' within one α-expansion of f
        if $E(f') < E(f)$ **then**
            $f = f'$
            success = true
        end if
    end for
until (success == false)

As mentioned above, the central part of the algorithm uses a single computation of the minimum cut to find the optimal α-expansion move.

Let us see how this is done and what is the structure of the associated graph.

The α-expansion computes a labeling corresponding to an *elementary cut* on a particular graph $G_\alpha(V_\alpha, A_\alpha)$. The structure of this graph is defined by the current partition P and by the label α. Note that the graph changes after each cycle of the algorithm.

The optimality is based on the fact that an elementary cut C on G_α is in one-to-one correspondence with labelings f^C that are within one α-expansion of the current labeling f and also that the cost of an elementary cut is $|C| = E(f^C)$. Consequently, the desired new labeling $\hat{f}$ is f^C, where C is a minimum cut on G_α.

The structure of the graph G_α is depicted in Figure 6.7. The set of vertices V consists of the vertices of the original graph $p \in P$ and the two

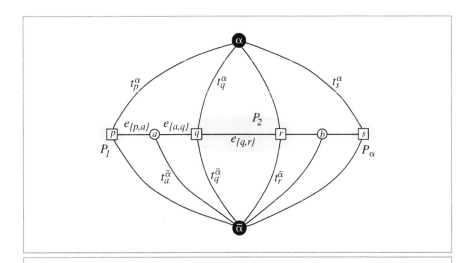

Figure 6.7: Graph G_α.

terminals α and $\bar{\alpha}$ (note that the label $\bar{\alpha}$ represents all other labels that are different compared with α). Additionally, a set of auxiliary vertices is created to take into account the boundaries in the current partition. That is, for each pair of neighboring vertices $\{p, q\}$ such that $f_p \neq f_q$, an auxiliary vertex $a_{\{p,q\}}$ is created.

$$V_\alpha = \left\{ \alpha, \bar{\alpha}, P, \bigcup_{\{p,q \in N | f_p \neq f_q\}} a_{\{p,q\}} \right\}.$$

In the graph, each original vertex $p \in P$ is connected to both terminals α and $\bar{\alpha}$ by t-edges t_p^α and $t_p^{\bar{\alpha}}$, respectively. Also, each pair of neighboring vertices $\{p, q\}$ with $f_p = f_q$ that are not separated by the current partition are linked by an n-edge $e_{\{p,q\}}$. For each pair of label-separated neighbor vertices $\{p, q\}$ with $f_p \neq f_q$, there is a triplet of edges $D_{\{p,q\}} = \{e_{\{p,a\}}, e_{\{a,q\}}, t_a^{\bar{\alpha}}\}$ that connect a to p, q, and $\bar{\alpha}$.

$$A_\alpha = \left\{ \bigcup \{t_p^\alpha, t_p^{\bar{\alpha}}\}, \bigcup D_{\{p,q\}}, \bigcup e_{\{p,q\}} \right\}.$$

The weights assigned to edges are given in Table 6.1.

A natural labeling f^C corresponding to a cut C on G_α is defined by

$$f_p^C = \begin{cases} \alpha & \text{if } t_p^\alpha \in C \\ f_p & \text{if } t_p^{\bar{\alpha}} \in C \end{cases} \qquad \forall p \in P. \tag{6.6}$$

In this way, a vertex p receives a label α if the cut C separates p from α. Vertices $p \notin P^\alpha$ maintain their previous label f_p.

Finally, to prove that the optimal α-expansion move generated by an elementary cut on G_α gives a labeling that is the local minimum of the energy function, all we need to do is to show that using a graph G_α constructed as above, for any elementary cut C, we have $|C| = E(f^C)$. That is, the cost of an elementary cut is equal to the lowest energy labeling.

Edge	Weight	For
t_p^α	∞	$p \in P_\alpha$
t_p^α	$E^i(f_p)$	$p \notin P_\alpha$
t_p^α	$E^i(\alpha)$	$p \in P$
$e_{\{p,a\}}$	$E^{i,j}(f_p, \alpha)$	
$e_{\{a,q\}}$	$E^{i,j}(\alpha, f_q)$	$\{p,q\} \in N, f_p \neq f_q$
$t_a^{\bar{\alpha}}$	$E^{i,j}(f_p, f_q)$	
$e_{\{p,q\}}$	$E^{i,j}(f_p, \alpha)$	$\{p,q\} \in N, f_p = f_q$

Table 6.1: Edge–weight assignments

The cost of an elementary cut is

$$|C| = \sum_{p \in P} |C \cup \{t_p^\alpha, t_p^{\bar{\alpha}}\}| + \sum_{\{p,q \in N | f_p = f_q\}} |C \cup e_{\{p,q\}}| + \sum_{\{p,q \in N | f_p \neq f_q\}} |C \cup D_{\{p,q\}}|.$$

(6.7)

By construction, for the edge–weight assignments of Table 6.1, the above expression is also equal to

$$|C| = \sum_{p \in P} E^i(f_p^C) + \sum_{p,q \in N} E^{i,j}(f_p^C, f_q^C) = E(f^C).$$

(6.8)

Therefore, the lowest energy labeling within an α-expansion-move from f is $\hat{f} = f^C$, where C is the minimum cut on G_α and $|C| = E(f^C)$.

6.7 BRANCH-AND-BOUND METHODS

Branch-and-bound methods are methods of intelligent enumeration for problems of combinatorial optimization.

These methods have two ingredients: branch and bound (establish thresholds). The strategy consists in estimating possible solutions and subdividing the problem.

6.7.1 CHARACTERISTICS

- Enumerates the alternatives, looking for branches based on estimates (bound) of the corresponding solution of an "easier" problem.
- Stops the enumeration in a branch when the estimate (upper bound) is inferior to the best known solution.
- Uses some heuristics in order to choose the node to be exploited: the one that has the greatest upper bound.
- Complexity: in the worst case, it is equal to the number of alternatives, and in practice, it is much smaller.

These methods are studied in more detail in Chapter 7.

6.8 APPLICATIONS IN COMPUTER GRAPHICS

In this section, we discuss the applications of combinatorial optimization the techniques in different areas of computer graphics. We focus on applications in image synthesis, geometric modeling, and vision.

Most of the applications of optimization methods in image synthesis are related to the problem of resource allocation. In general, the visualization system has limited processing or memory resources for exhibiting images. Therefore, it is necessary to use computational techniques to adapt the generated images to the available resources. More precisely, we consider two examples. The first one is image quantization, and the second one is interactive visualization using levels of details.

Most of the applications of optimization methods in geometric modeling are related to the problem of constructing optimal geometric shapes. In general, a modeling system integrates several tools in order to create and analyze objects. In the applications, the shape of the objects must comply with certain functional or aesthetic requirements. For example, a curve must pass through some predetermined points and, at the same time, it must be smooth. For this reason, normally, it is very difficult for the user to accomplish such tasks manually, without using appropriate tools. Optimization methods enable the use of a natural formulation of the problem and its automatic solution. We study two examples of the use of combinatorial optimization methods in geometric modeling. The first example is the computation of curves of minimum length in maps; the second example is the reconstruction of surfaces from a finite set of transversal sections.

Most applications of optimization methods in computer vision are related to classification and inference problems. In particular, many low-level vision problems can be posed as a *pixel-labeling* problem, where different formulations are related to the model to be estimated, such as in stereo correspondence and image segmentation.

We study examples of the use of graph cuts to solve the pixel-labeling problem. The example is in two-view stereo correspondence.

6.8.1 IMAGE QUANTIZATION

The problem of image quantization is a good example of a problem that exploits the idea of limited resources in order to model it as a combinatorial optimization problem.

The problem of quantization can be stated as follows:

Find a set of k colors that represent the gamut of an image f(u, v) with the smallest distortion.

The problem is clearly an optimization problem. Moreover, since a digital image has a discrete representation, the problem can be posed as a combinatorial problem.

In order to better formulate the problem, we need some preliminary definitions.

An *image* is a graphical object $\mathcal{O} = (U, f)$, with $f\colon U \subset \mathbb{R}^2 \to C$, where C is a color space. The attribute function f is called *image function*. The set U is called the *image support*, and the image set $f(U)$ is called the *image gamut*.

In graphical applications, the function f describes a digital image, which has a discrete representation. In this way, in general, both the domain and the range set the attribute function f are discretized and, for this reason, f is called a discrete–discrete image. The image support is usually a rectangle $U = [a, b] \times [c, d]$ sampled according to a uniform grid $F_\Delta = \{(u_i, v_j) \in \mathbb{R}^2\}$, where $u_i = a + i\Delta u$ and $v_j = c + j\Delta v$, with $\Delta u = (b - a)/m$, $\Delta v = (d - c)/n$, and $i = 0, \ldots m, j = 0, \ldots n$. The points (u_i, v_j) are called sample points.

The color space C of the image is usually a trichromatic color space ($C \equiv \mathbb{R}^3$) or a monochromatic one ($C \equiv \mathbb{R}$). Moreover, the values of f in the sample points are usually represented by an integer number with l bits (another option would be the representation by a floating-point number). That is, the image gamut is a finite set with 2^l elements.

Figure 6.8 shows a discrete–discrete monochromatic image with its domain U and range set C.

Quantization

A quantization of an image is the process of discretization or reduction of its color gamut. Now we define in a more precise way the problem of quantization. A *quantization* is a subjective transformation $q\colon \mathbb{R}^n \to M_k$, where $M_k = \{c_1, c_2, \ldots c_k\}$ is a finite subset of $\mathbb{R}^n$. The set M_k is called *color map* of the quantization transformation. When $k = 2^l$, we say that q is a quantization with l bits. Given a discrete representation of an image, it is usual to have a quantization among finite subsets of

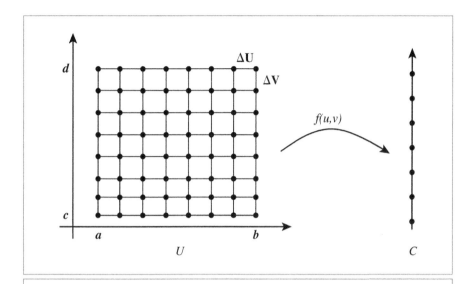

Figure 6.8: Monochromatic discrete image.

color, of the type $q\colon R_j \to M_k$. If $j = 2^n$ and $k = 2^m$. In this case, we say that we have a quantization from n to m bits.

Consider a quantization map $q\colon \mathbb{R}^n \to M_k$. The elements c_i of M_k are called *quantization levels*. To each quantization level $c_i \in M_k$, there corresponds a subset of colors $C_i \subset \mathbb{R}^n$, which are the colors mapped on to the color c_i by the transformation q, i.e.,

$$C_i = q^{-1}(c_i) = \{c \in C; q(c) = c_i\}$$

The family of sets C_i constitute a partition of the color space. Each of the partition sets C_i is called *quantization cell*. Note that the quantization map q assumes a constant value, equal to c_i, on each cell C_i.

When the color space is monochromatic ($C \equiv \mathbb{R}$), we say that the quantization is one dimensional. We concentrate here in this simpler case, although most of the concepts we introduce extend naturally to the case of color images. Figure 6.9 shows the quantization levels and the graph of a ID quantization function.

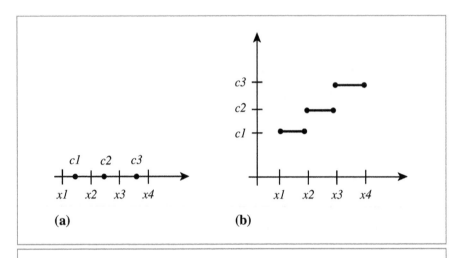

Figure 6.9: Quantization levels (a) and graph of an ID quantization map (b).

Note that the quantization function q is completely determined by the quantization cells C_i and the quantization levels c_i. In fact, using geometric arguments, we can obtain c_i from C_i and vice versa. Thus, some quantization methods compute first the levels c_i and then the cells C_i, while others use the inverse strategy.

In order to estimate the effectiveness of the quantization transformation, we have to define the notion of *quantization error*. Let c be a color to be quantized and q the quantization map, then $c = q(c) + e_q$. The quantization error $e_q = c - q(c)$ is the distance $d(c, q(c))$ between the original color c and the quantized color $q(c)$, using an appropriate metric. The metric root mean square, given by the square of the Euclidean norm, is widely used because of its simplicity.

The quantization error associated with a cell C_k is given by

$$E(k) = \sum_{c \in C_k} h(c)d(c, c_k), \qquad (6.9)$$

where $h(c)$ is the number of occurrences of the color c in the image f. This number is given by the frequency *histogram* of the image that provides for each color the number of pixels that is mapped to that color (note that the normalized histogram is an approximation to the probability distribution of the colors of the image f).

Given a quantization cell C_k, it is possible to prove that the optimal level of quantization c_k corresponding to C_k is the centroid of the set of colors $\{c; c \in C_k\}$ weighted by the distribution of the colors in C_k.

The total quantization error for an image f is the sum of the quantization error over all of the samples of the image

$$E(f, q) = \sum_i \sum_j d(f(u_i, v_j), q(f(u_i, v_j))). \qquad (6.10)$$

We remark that the above equation can be expressed using the quantization cells and the levels of quantization as $E(f, q) = \sum_k E(C_k)$.

The quantization problem consists in computing c_i and C_i in such a way that the error $E(f, q)$ be minimum. This is equivalent to obtaining the optimal quantizer $\hat{q}$ over all the possible K partitions Q_K of the color space

$$\hat{q} = \arg \min_{q \in Q_K} E(f, q). \qquad (6.11)$$

Quantization Using Dynamic Programming

We have seen that the problem of color quantization can be stated as an optimization problem over a discrete set. Nevertheless, the size of the solution set of the Equation 6.11 is of order $|Q_K| = \binom{N-1}{K-1} = O(N^{K-1})$, where K is the number of quantization levels and N is the number of distinct colors of the image f.

Clearly, we need an efficient optimization method to solve the problem. We will show how to use dynamic programming for the case of monochromatic quantization ($C \equiv \mathbb{R}$).

Let q_k^n be the optimal quantizer of n to k levels. The first t cells of q_k^n, $1 < t < k$ with $k < n$, constitute a partition of the interval $[0, x_t]$ of gray levels. It is possible to prove that these cells constitute an optimum quantizer for t levels of the subset of the colors $c \in [0, x_t]$. This property is equivalent to the principle of optimality and allows us to obtain a solution to the problem using dynamic programming.

Let $L(n, k) = x_{k-1}$, the superior limit of the $(k-1)$-th cell of the optimal quantizer $\hat{q}_k^n$. Then, by the principle of optimality,

$$L(n, k) = \arg \min_{x_k < x_i < x_n} \{E(f, \hat{q}_{k-1}^i) + \mathcal{E}(x_i, x_n)\}, \qquad (6.12)$$

where $2 \leq k < n \leq N$, and $\mathcal{E}(a, b)$ is the total quantization error of the interval $[a, b]$. Note that we can determine $L(n, k)$ by using linear search if we know $E(f, \hat{q}_{k-1}^i)$, for $k \leq i < n$. Computing $L(n, k)$, we can compute $E(f, \hat{q}_k^n)$

$$E(f, \hat{q}_k^n) = E(f, \hat{q}_{k-1}^{L(n,k)}) + \mathcal{E}(L(n, k), x_n). \qquad (6.13)$$

The basic idea is to use a recursive process. Given the interval $[x_0, x_N)$, we compute the first optimum quantization cell $C_1 = [x_0, x_1)$ of this interval and we repeat the process for the complementary interval $[x_1, x_N)$ until we obtain the K cells of q_K^N.

Note that the errors of the quantizers of level 1 are computed trivially $E(f, q_1^n) = \mathcal{E}(0, x_n)$, with $1 < x_n < x_N$. Moreover, using (6.12) and (6.13), it is possible to compute $L(n, 2)$ and $E(f, q_2^n)$, $2 \leq n \leq N$ and these values are to be used in the recursion of the Algorithm 6.

Algorithm 6 Quantize (f, N, K)

$E[i] = \mathcal{E}(0, x_i)$, for $1 \leq i \leq N$
$L[k, k] = k - 1$, for $1 \leq k \leq K$
for $k = 2, \ldots, K$ **do**
  for $n = k + 1, \ldots, N - K + k$ **do**
    $x = n - 1$
    $e = E[n - 1]$
    for $t = n - 2, \ldots, k - 1$ **do**
      if $E[t] + \mathcal{E}(x_t, x_n) < e$ **then**
        $x = t$
        $e = E[t] + \mathcal{E}(x_t, x_n)$
      end if
    end for
    $L[n, k] = x$
    $E[n] = e$
  end for
end for
return $q_K^N = (L[x_k, k])_{k=1,\ldots,K}$

In Figure 6.10, we show the result of applying this algorithm to obtain the quantization of an image from 256 to 4 gray levels.

Final Remarks

The Algorithm 6 depends strongly on the fact that in the monochromatic quantization, the cells are intervals, and we have a natural order of the color space partition. This is the basis for the sequencing of the stages of the dynamic programming.

This algorithm can be extended to color quantization if we establish an ordering of the quantization cells. One way to obtain this is to consider the projection of the colors of the image along a direction and create quantization cells with orthogonal planes to this direction. This method works well because, in general, the colors of an image are concentrated along a principal direction.

(a)

(b)

Figure 6.10: Quantization from 256 (a) to 4 (b) gray levels.

Both versions, monochromatic and polychromatic, of this algorithm were introduced by Wu (1992).

6.8.2 INTERACTIVE VISUALIZATION

Now we cover a problem in the area of visualization, which is in the category of the use of limited resources in graphics processing.

The main objective of the interactive visualization is to exhibit data, which can vary dynamically, in such a way to allow their exploration and manipulation by the user.

For this purpose, images of the data must be generated at interactive rates, i.e., in a way to make it possible a feedback of the process by the user based on image updating.

This type of demand exists in every interactive graphical applications, such as geometric modeling systems and CAD systems.

A stronger constraint is the visualization in real time implemented in flight simulators and other similar applications. In this case, besides the time to drawing the image being shorter (typically 1/60 of a second), no variation in the exhibition rate is allowed.

Based on the above description, the interactive visualization problem of data can be stated as follows. *We should produce the best visual representation of the data within the limits of the allocated time interval.*

This informal definition of interactive visualization makes it clear that this is an optimization problem (produce the best image) with constraint (allocated time).

We concentrate our efforts in the visualization of 3D scenes, although most of the concepts can be applied in the visualization of other types of data, such as 2D scenes, sequence of images (animation), or even volumetric data.

A 3D scene is composed of a set of graphical objects positioned in relation to a common referential, called global coordinate system (or scene coordinate system).

A graphical object $\mathcal{O} = (U, f)$ consists of its geometric support $U \subset \mathbb{R}^3$ and its attributes $f: U \to \mathbb{R}^p$. The geometry of a graphical object defines its shape. The attributes of a graphical object define its properties, such as color, reflectance, which, in general, are related to its visual appearance.

Probably the most common example is a 2D polygonal object embedded in the 3D space. The shape is given by a surface described by a mesh of triangles, and its attributes are defined on the vertices of this mesh.

A graphical object of *variable resolution* incorporates in the representation the possibility of extracting approximate versions of its geometry. These are simplified representations, also called *levels of details*, which can be attained to satisfy two different types of criteria: *approximation error* (local or global) and *size of the representation*.

In Figure 6.11, we show a polygonal graphical object using three levels of detail,with 32, 428, and 2267 polygons.

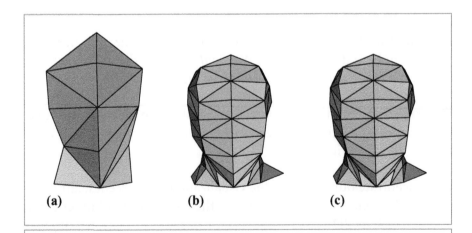

(a) **(b)** **(c)**

Figure 6.11: Spock represented with three levels of detail.

A graphical object can be composed of several subgraphical objects. Moreover, these objects can be structured in a hierarchical form. In the first case, we have a *composed graphical object*, and in the second case, we have a *hierarchy of graphical objects*.

In general, it is possible to substitute a group of graphical objects with a unique graphical object that represents the group. This operation is called *clustering*. The resulting graphical object is called *impostor*.

The hierarchical representation is constituted by collections of graphical objects clustered according to logical levels. At each level change, a group of subordinated objects (sons) may have their representation substituted with a unique simpler graphical object (father). This type of hierarchy is called *level of detail hierarchy*.

Some graphical objects have a natural hierarchy. This is the case of objects with articulated structures. For example, the human body can be represented by the following hierarchy:

$$
\text{Body}
\begin{cases}
\text{Head} \\
\text{Torso} \\
\text{Right Arm} \begin{cases} \text{Upper Arm} \\ \text{Lower Arm} \end{cases} \\
\text{Left Arm} \begin{cases} \text{Upper Arm} \\ \text{Lower Arm} \end{cases} \\
\text{Right Leg} \begin{cases} \text{Thigh} \\ \text{Lower Leg} \end{cases} \\
\text{Left Leg} \begin{cases} \text{Thigh} \\ \text{Lower Leg.} \end{cases}
\end{cases}
$$

For graphical objects with hierarchical structure, the creation of a hierarchy of levels of details is easy. In Figure 6.12, we show the

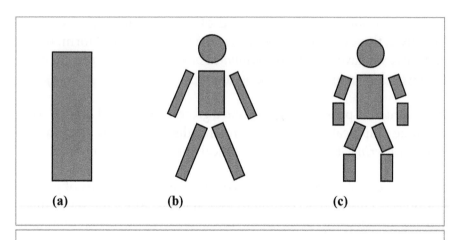

Figure 6.12: Human body with three levels of clustering.

representations corresponding to the three levels of clustering of the hierarchy of the human body.

We should remark that the clustering operation can be attained in arbitrary sets of graphical objects in a 3D scene, even if these objects do not have a natural hierarchical structure. In this case, the clustering is done based on the proximity relations between the objects. These groups can be grouped to constitute a *hierarchy of the scene objects.*

Optimization of Level of Details

Interactive visualization implies the maintenance of a fixed rate in the generation of images from the scene. That is, the time to depict the objects of the scene on the screen cannot be greater than a predefined threshold. We have seen that the multiresolution representation of graphical objects allows us to control their geometric complexity and, consequently, the time to draw them.

Interactive visualization methods need to determine the level of detail of the objects in the scene in such a way that the time to draw is below the threshold.

There are two types of strategy for this purpose: reactive and predictive methods. *Reactive methods* are based on the time spent drawing previous images. If this time is greater than the threshold, the level of detail is reduced; otherwise, it is increased. *Predictive methods* are based on the complexity of the scene to be visualized. Thus, these methods need to estimate the drawing time corresponding to each level of detail of the objects.

Although predictive methods are more difficult to implement, they are the only ones that allow us to guarantee a fixed rate of image generation. For this reason, we develop a predictive method using optimization.

Before describing the method, we have to introduce some definitions.

A *visual realization* of the graphical object $\mathcal{O}$, determined by the triple $(\mathcal{O}, L, R)$, corresponds to a unique instance of $\mathcal{O}$ in the level of detail L, which has been drawn using the algorithm R.

We define two heuristics: $\text{cost}(\mathcal{O}, L, R)$ and $\text{benefit}(\mathcal{O}, L, R)$. The cost function estimates the time necessary to generate a realization of the object. The benefit function estimates the contribution of the object realization for the perceptual effectiveness of the scene. We also define the set V of objects of the scene, which are visualized in the current frame.

With these definitions, we can pose the interactive visualization problem as

$$\text{maximize} \sum_{(\mathcal{O}_i, L_i, R_i) \in C} \text{benefit}(\mathcal{O}_i, L_i, R_i) \qquad (6.14)$$

$$\text{subject to} \sum_{(\mathcal{O}_i, L_i, R_i) \in C} \text{cost}(\mathcal{O}_i, L_i, R_i) \leq T, \qquad (6.15)$$

where T is the time allowed to draw the image.

We should remember that this formulation, besides capturing the essential aspects of the problem, can be applied in different contexts related to image synthesis. This depends on the type of geometric representation and of the class of algorithms used.

Applying the above method requires that we have an efficient algorithm to compute the functions cost and benefit. We assume that this is possible for the practical cases.

We describe now some heuristics to estimate these functions in the case of multiresolution polygonal meshes using the z-buffer algorithm.

The cost function depends on two factors:

- *primitive processing*: coordinate transformations, clipping, illumination.
- *pixel processing*: rasterization, z-buffer, texture mapping, Gouraud interpolation.

These factors correspond to the two stages of the visualization process (see Figure 6.13).

Using this model, we estimate drawing time by

$$
\text{cost}(\mathcal{O}, L, R) = \max \begin{cases} c_1 R_{\text{poly}}(\mathcal{O}, L) + c_2 R_{\text{vert}}(\mathcal{O}, L) \\ c_3 R_{\text{pix}}(\mathcal{O}) \end{cases},
$$

where R_{poly}, R_{vert}, and R_{pix} are, respectively, the processing times of polygons, vertices, and pixels of the algorithm R, and c_1, c_2, and c_3 are constants that depend on the hardware used (i.e., FLOPS).

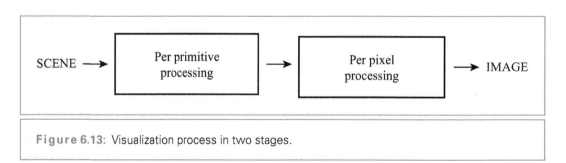

Figure 6.13: Visualization process in two stages.

The function benefit depends mainly on the *area* occupied by the object and on the *fidelity* of its visualization. Moreover, this function depends on several perceptual factors:

- *semantics*: define the relative importance of the objects
- *focus*: given by the region of interest
- *speed*: depends on the motion (motion blur)
- *hysteresis*: level of detail must vary smoothly.

Based on these elements, we estimate the effectiveness of the image as

$$\text{benefit}(\mathcal{O}, L, R) = \text{area}(\mathcal{O}) \times \text{fidelity}(\mathcal{O}, R) \times$$
$$\text{importance}(\mathcal{O}) \times \text{focus}(\mathcal{O}) \times$$
$$\text{mov}(\mathcal{O}) \times \text{hyst}(\mathcal{O}, L, R).$$

The reader should observe that the preponderant factors in the above equation are the size of the object (i.e., area($\mathcal{O}$)) and the fidelity of its realization, which is determined by the reconstruction used in the visualization algorithm.

The interactive visualization method by optimization uses the cost and benefit functions defined above in order to choose the representation set C that satisfies the Equation 6.14 for each frame.

The reader should observe that this optimization problem with constraints is a version of the knapsack problem in which the items are partitioned into subsets of candidates and only one item of each subset can be inserted in the knapsack. In this case, the items in the knapsack constitute the set C, and the realizations $(\mathcal{O}_i, L_i, R_i)$ are the items inserted in the knapsack of each subset S_i.

Unfortunately, this version of the knapsack problem is *NP*-hard. However, a reasonable solution is obtained using a greedy algorithm. This algorithm, at each step, chooses the realization of greatest value v, given by the ratio

$$v = \frac{\text{benefit}(\mathcal{O}_i, L_i, R_i)}{\text{cost}(\mathcal{O}_i, L_i, R_i)}.$$

The items are selected in decreasing order of value until it fills the size of the knapsack, given by the Equation 6.15. We remark that since only a realization of each object $\mathcal{O}_i$ can be part of the set C, its realization $(\mathcal{O}_i, L_i, R_i)$ with greatest value v will be added to the knapsack.

Final Remarks

The interactive visualization by optimization was developed by Funkhouser and Séquin (1993) for the case of multiple levels of detail without clustering. The problem was also studied by Mason and Blake (1997), who extended the method to the case of hierarchical representation of the scene, with clustering.

6.8.3 SHORTEST PATHS

In some modeling applications, it is necessary to compute curves of minimum length on a surface. This problem is specially relevant in GIS applications. In this type of application, we need to find shortest paths in maps for different purposes, such as in the planning to build a highway.

Combinatorial optimization methods allow us to compute minimal curves in general graphs. We show how to reduce the problem of finding optimal paths in maps to the problem of shortest paths in neighborhood graphs.

In GIS, we treat data defined on the earth's surface. A map is a georeferenced region that can be modeled as a 2D graphical object $\mathcal{O} = (U, f)$, where the geometric support U is a rectangle $U = [a, b] \times [b, c]$ and the attribute function f describes geographic information in U.

Depending on the nature of the attributes, we have two types of geographic objects:

- entity maps
- property maps.

In entity maps or *thematic maps*, f defines a partition of the region U into subregions S_i, with $i = 1, \ldots, N$, each one being associated with a geographic entity E_i. For example, a geopolitic map, with the division of a country into its states, is in this category.

In property maps, also called *field maps*, f defines continuous attributes that vary on the region U. For example, an elevation map, describing the topography of a region, is in this category.

We remark that for a given region U, we might have several distinct attribute functions f_U associated with it. This is equivalent to saying that the range set of the attribute function is multidimensional.

The fundamental operations in a GIS correspond to the analysis of these geographic objects. In this context, the relevant problems are related to

1. the data correlation between different attribute functions
2. the inference of new attribute functions from basic data.

The problem of shortest curves in maps is in this last class of operations.

Until now, we have discussed a continuous mathematical model for geographic objects. The computational implementation requires a discrete representation of this model.

The discretization of a geographic object is, in general, based on a matrix or a vector representation. The vector representation is more appropriate to describe entity maps, while the matrix representation is the choice to describe property maps.

Example 24 [satellite images]. Remote sensing satellites capture the irradiation from regions on the earth's surface. These data are discretized by the sensors and are structured in such a way that they are ready for a matrix representation. Therefore, we have a discrete image on a regular grid.

From the matrix data, such as satellite images, we are able to extract other information. For example, the terrain elevation can be computed from the correlation of points in two images. This operation is known as aerophotogrametry and is related to the problem of stereographic vision, which we discuss later.

Map operations can be defined by n variable functions (combination functions), or by functions of one variable (transformation functions). For example, we can define a transformation operation to compute the inclination of a terrain from its topographic map.

The shortest path problem consists in finding a minimal length curve in U that joins the point $p_0 \in U$ to the point $p_1 \in U$. The metric used is defined by an attribute function of U. This problem, in its discrete form, can be solved using combinatorial optimization methods.

Shortest Paths in Neighborhood Graphs

In this section, we pose the discrete problem of optimal paths in maps. The input data consist of an image represented by a matrix (raster image) C, the initial point p_0, and final point p_1 of the path to be computed, both points belonging to C.

The image C is given by a matrix $M \times N$, in which the value $c(i, j) \in \mathbb{R}^+$ of entry (i, j) provides the cost to traverse the pixel (we can interpret this value as the diameter of the pixel using the metric defined by C).

The problem consists in computing the path of minimum cost joining the points $p_0 = (i_0, j_0)$ and $p_1 = (i_1, j_1)$.

The first step to solve the problem is to construct an adequate graph. We have two parameters to consider: the density of the graph and the neighborhood of each graph node.

We define the neighborhood graph $G = (V, A)$, where V is the set of graph vertices and A is the set of arcs that join two neighbor vertices.

The set V is a discretization of the input image C. For convenience, we use the original discretization of C. Thus, V is the set of the $M \times N$ pixels $(i, j) \in C$, where $i = 0, \ldots, N - 1$ and $j = 0, \ldots, M - 1$.

The set A is determined by a topology on the grid c_{ij}, from which we define the discrete neighborhood of a pixel. Two natural options are the four-connected and eight-connected neighborhoods (see Figure 6.14).

We use the topology defined by the four-connected neighborhood. Thus, for each pixel (i, j), we construct four arcs,

$$
\begin{aligned}
a_l &= ((i, j), (i + 1, j)) \\
a_r &= ((i, j), (i - 1, j)) \\
a_t &= ((i, j), (i, j + 1)) \\
a_b &= ((i, j), (i, j - 1)),
\end{aligned}
$$

that join (i, j) to its neighbors $(i + 1, j)$, of the left $(i - 1, j)$, from above $(i, j + 1)$ and from below $(i, j - 1)$.

To each arc $a = ((i, j), (k, l))$, we associate a cost,

$$
c_a((i, j), (k, l)) = \frac{c(i, j) + c(k, l)}{2}.
$$

With the problem structured in this form, we can solve it using Dijkstra algorithm because the costs are all positive. We remark that it is not necessary to construct the graph directly to obtain the path.

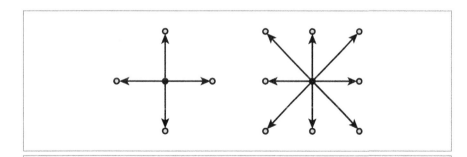

Figure 6.14: Neighborhoods of a pixel.

Final Remarks

The algorithm Voxel Coding (Zhou *et al.,* 1998) solves a simplified version of the problem for *n*-dimensional volumetric data with functions of constant cost.

The shortest path problem is related to the computation of geodesic curves on surfaces. The main difference is that in the case of shortest paths, the curve is constrained to go through the edges of the mesh that discretize the surface, while in the case of geodesics, the curve can pass anywhere in the domain.

6.8.4 SURFACE RECONSTRUCTION

The reconstruction of 3D surfaces from plane curves is a powerful technique in geometric modeling. This technique has special relevance in several computer graphics application areas. In particular, it is used in the area of medical images to create anatomic models from volumetric data. This is the case of CT and MRI. Moreover, the reconstruction of surfaces using transversal sections is closely related to the *lofting* technique, which is widely used in CAD/Computer Aided Manufacturing, which allows us to construct solid shapes using the extrusion of contours.

The surface reconstruction problem using transversal sections consists in producing the best surface that contains a set of predefined curves.

This problem is clearly an optimization one, and when the curves have a discrete description, it can be solved using combinatorial optimization methods.

Some Definitions

The data for the problem of surface reconstruction by transversal sections consist in a set of transversal sections $\{S_i | i = 1, \ldots, K\}$ defined in planes orthogonal to a given direction z, corresponding to a monotone sequence (z_i), $i = 1, \ldots, K$ (i.e., $i < i + 1, \forall i$).

Each transversal section S_i consists of one or more contours, given by polygonal closed curves $P = (p_0, p_1, \ldots p_n)$ (i.e., $p_0 = p_n$) that represent the intersection of the surface to be reconstructed with the support plane z_i of the section S_i. In order to simplify the problem, we assume that each section has only one contour, i.e., each contour curve is connected.

The discrete reconstruction problem consists in finding the mesh of triangles that contains the given polygonal curves and at the same time satisfies certain optimality criteria.

We should remark that it is possible to decompose the above problem into a sequence of simpler subproblems: the construction of meshes determined by pairs of contours in adjacent transversal sections S_i and S_{i+1}.

A natural interpretation in this context is to see the problem as the problem of finding the topology of the triangle mesh that corresponds to the polygonal surface with "best" geometry (we remark that the definition of the optimality criteria depends on the precise definition of a "good" surface).

In order to construct the surface topology and geometry defined by consecutive curves, we have to construct the polygons between these two contours. Essentially, the problem is reduced to the one of finding vertex correspondences in the contours in order to construct the triangles.

Optimization Methods for Contour Reconstruction

The use of combinatorial optimization methods in the solution to the problem requires that the topological relations be formulated using a graph.

We define the contours $P \in S_i$ and $Q \in S_{i+1}$ by vertex lists $P = (p_0, p_1, \ldots p_{m-1})$ and $Q = (q_0, q_1, \ldots q_{n-1})$, with $p_k, q_l \in \mathbb{R}^3$. Since the contours represent closed polygonal curves, p_0 follows p_{m-1} and q_0 follows q_{n-1}. That is, the indices of P and Q must be interpreted modulo m and n, respectively.

Because of the nature of the problem, the valid triangles of the mesh are *elementary triangles*, for which one of the edges belongs to one of the contours (P or Q) and the other two edges join this edge to a vertex in the other contour. That is, an elementary triangle is of the form (q_j, q_{j+1}, p_i) or (p_{i+1}, p_i, q_j) (see Figure 6.15).

In an elementary triangle, we distinguish two types of edges: we call *segments* the edges that belong to the contours P and Q; we call *span* the edges that join one contour to the other.

A polygonal mesh that reconstructs the region of the surface between the contours P and Q has the following properties:

$M1$—each segment is the edge of exactly one triangle of the mesh
$M2$—if there exists a determined span, it is the common edge of two triangles adjacent to the mesh.

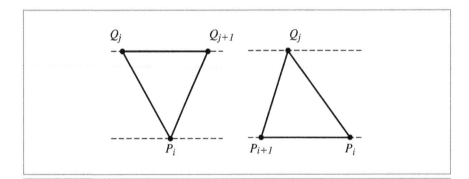

Figure 6.15: Elementary triangle.

The surface defined by this mesh is called *acceptable surface.*

There exists a great number of meshes that satisfy the requirements, $M1$ and $M2$ above. In order to better understand the valid meshes, we define the directed graph $G = (V, A)$ in which the nodes $v \in V$ correspond to the set of all the possible spans between the vertices of the contours P and Q, and the arcs $a \in A$ correspond to the set of elementary triangles of the mesh. More precisely,

$$V = \{v = (i, j) : v \mapsto (p_i, q_j)\}$$

$$A = \{a = (i, j, k, l) : a \mapsto (p_{i+1}, p_i, q_j), a \mapsto (q_j, q_{j+1}, p_k)\}.$$

We remark that an edge $a = (v_{i,j}, v_{k,l})$ represents a triangle delimited by the spans (p_i, q_j) and (p_k, q_l). The other edge, given by (p_i, p_k) or (q_j, q_l), belongs to the contour P or Q. Thus, elementary triangles satisfy one of the constraints: $i = k, l = j + 1$, or $k = i + 1, l = j$.

Since the polygonal curves are closed, G is a *toroidal* graph (see Figure 6.16).

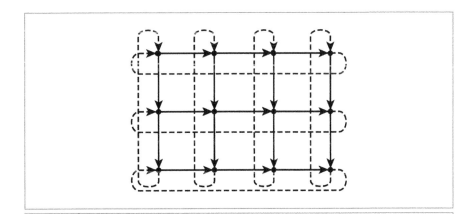

Figure 6.16: Toroidal graph.

A valid mesh corresponds to a cycle S in the graph G. The properties $M1$ and $M2$ of an acceptable surface can be interpreted based on this graph in the following form:

$G1$—for each index $i = 0, \ldots, m - 1$, there exists exactly a unique arc $(v_{i,j}, v_{k,j+1}) \in S$.

$G2$—if the vertex $v_{i,j} \in S$ exists, then exactly one arc $a_I \in S$ ends at $v_{i,j}$, and an arc a_O departs from $v_{i,j}$.

From the above remarks, we conclude that the number of valid meshes increases exponentially with the number of vertices in the graph (mn).

It is our goal to find a valid mesh that represents the best surface whose boundary is defined by the contours P and Q. Thus, we have formulated the reconstruction problem by contours as a combinatorial optimization problem.

"Find the valid cycle of minimal cost in graph G."

For this, we have to choose an appropriate metric that will be used to provide quantitative information about the quality of the mesh. A local metric associates with each arc $(v_{i,j}, v_{k,l}) \in A$ a cost $c : A \to \mathbb{R}^+$. The function $c(v_{i,j}, v_{k,l})$ estimates the triangle contribution corresponding to the quality of the mesh.

There are several possible local metrics:

- **volume**—maximizing the volume is a good metric for convex objects; nevertheless, it does not perform well for concave objects. Moreover, this metric is very difficult to compute.
- **edge length**—minimizing the length of spans favors the construction of ruled surfaces. The computation of this metric is simple.
- **area**—minimizing the total area of the surface is, probably, the most general metric. It works very well for concave and convex surfaces and is easy to compute.

We remark that we can define variants of the above metrics. For example, normalize the pairs of contours such that both are mapped into a unit-evolving box.

Back to our optimization problem, we need to compute the cycle S of minimum cost according to the chosen metric and that satisfies properties $G1$ and $G2$.

The solution to the problem can be considerably simplified by using an appropriate transformation in the graph G that reflects the properties of the valid cycles. We define the graph G' obtained from gluing two copies of G. More precisely, $G' = (V', A')$, where

$$V' = \{v_{i,j} : i = 0, \ldots, 2m; j = 0, \ldots, n\}$$

and A' is defined in, similarly, analogous to A.

The cost c' of an edge in G' is given by

$$c'(v_{i,j}, v_{k,l}) = c(v_{i\bmod m, j\bmod n}, v_{k\bmod m, l\bmod n}).$$

Note that, contrary to G, the transformed graph G' is acyclic.

With this new formulation of the problem, we conclude that the set of valid cycles in G is given by the paths $\gamma_k \in G'$ that have $v_{k,0}$ e $v_{m+k,n}$, respectively, as initial and final vertices. To recover the cycle $S_k \in G$ equivalent to $\gamma_k \in G'$, we just need to take the coordinates $v_{i,j}$ as $v_{i\bmod m, j\bmod n}$.

In this way, the solution to the problem is given by the path γ_k of smallest cost for $k \in (0, \ldots, m - 1)$.

$$\gamma_{\min} = \min_{k=0, m-1} \text{OptimalPath}(k, m + k),$$

where the routine OptimalPath computes the shortest path joining the vertices $v_{k,0}$ and $v_{m+k,m}$ in G'.

Note that since all the costs are positive, it is also possible in this case, use Djkstra algorithm, to implement the routine OptimalPath.

Final Remarks

A more efficient version of the combinatorial optimization algorithm described above uses the *divide-and-conquer* strategy. It exploits the fact that if two optimal paths cross themselves, they must have a part in common (Fuchs *et al.*, 1977).

We remark that the problem of surface reconstruction from cuts is more general than we have discussed in this section.

In the general case, there may exist several contours for each section (i.e., the contour is not connected), and the complete problem is divided into the following subproblems (Meyers *et al.*, 1991):

- **correspondence**—find correspondences between contours in each pair of adjacent sections
- **bifurcation**—when a unique contour in a section corresponds to more than one contour in another section, this section must be subdivided
- **polygonization**—find the polygonal mesh between two contours on adjacent sections (this problem has been discussed here).

Another interesting remark is that the polygonal mesh produced by the reconstruction algorithm can be coded in a compact form using a *triangle strip*.

6.8.5 STEREO

Stereo reconstruction is one of the fundamental problems in computer vision. The goal is to recover 3D information from a pair of 2D images of a 3D scene. In a broad sense, this problem is the machine vision equivalent of the binocular natural vision perception that allows humans to infer depth information of real-world scenes based on the images from both eyes.

Stereo reconstruction has applications in many areas, such as robotic navigation and geometric modeling.

The Geometry of Stereo Reconstruction

The basic geometric principle of 3D reconstruction from an image pair is known as *triangulation*. In this setting, it is assumed that the images come from cameras that have been *calibrated* (as presented in Section 5.6.1). That is, the intrinsic and extrisic parameters of the two cameras are known, as well as their relative geometric position.

Note that any visible 3D point X in the scene is projected into 2D points x and x' in each of the images. In stereo reconstruction, the unknowns are the 3D coordinates of scene points, and the known information consists of the corresponding image projections and the camera information. Thus, the reconstruction problem can be formulated as follows. Given two corresponding points $x \in I$ and $x' \in I'$ in images I and I', as well as the camera calibration data, find the 3D point X whose image is projected to x and x'.

The geometry of the problem is depicted in Figure 6.17. For each of the two cameras, we have their centers of projection c and c' with associated coordinate systems. We are given also the corresponding image points x and X. The triangulation principle is based on the fact that from these parameters, we can compute two view rays r and r', one for each camera, such that intersect is exactly at the 3D point X. The construction is simply done by taking $r(t) = c + t(x - c)$, where c is the origin of the ray and the vector $(x - c)$ is the direction of the ray. Similarly for the other ray, $r'(t) = c' + t(x' - c')$. We remark that the term triangulation is used here because the points X, c, and c' are the vertices of a triangle, as can be observed in Figure 6.17.

The description above may give the wrong impression that triangulation is all it is required for 3D stereo reconstruction. On the contrary,

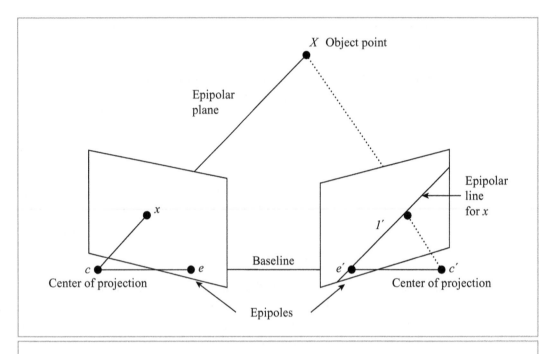

Figure 6.17: Stereo image pair and epipole geometry.

in a certain sense, triangulation is the easy part of the problem. The perceptive reader may have noticed that in the statement of the triangulation problem, we assumed known the correspondence between image points x and x'. However, this knowledge may be automatic for a human observer who has a high-level understanding of the scene but is very difficult for a machine to compute just from raw images. So, in order to perform triangulation, it is necessary first to compute *correspondences* of points in the two images.

Summarizing, we can say that the problem of 3D reconstruction is divided into three subproblems: calibration, correspondence, and triangulation. Arguably, the hardest one is the correspondence problem.

The stereo correspondence involves recovering matching pixels corresponding to the same 3D feature in the two views of the scene.

Finding matching pairs for all pixels in the two images is difficult to solve accurately because such match can be highly ambiguous. This inherent ambiguity comes from many factors, such as image noise, specularities of objects, and lack of texture in the scene. The latter factor of uncertainty is well known in computer vision under the name of *window* effect.

In a stereo pair, the difference between the image coordinates of matching pixels is called *disparity*. The disparity is proportional to the depth of the associated 3D point relative to the cameras. Also, pixels in one image and their disparity determine the correspondence. Conversely, the correspondence problem can be posed as finding the disparity associated with each pixel in the image.

At first, it might appear that the correspondence problem requires a 2D search on the neighborhood of a given pixel in the other image. But the *epipolar constraint* reduces this search to a single line, which simplifies the problem considerably.

To understand the epipolar constraint, let us give some definitions. The *baseline* is the line joining the centers of projections c and c'. The point of intersection of the baseline with the image plane is called *epipole*. Thus, the epipole is the image, in one camera, of the other camera's center of projection. The *epipolar plane* is the plane defined by a 3D point X and the centers of projection c and c'. The *epipolar line* is the straight line of intersection of the epipolar plane with the image plane. It is the image in one camera of a ray through the optical center and image point in the other camera.

All epipolar lines intersect at the epipole (see Figure 6.17). Therefore, a point x in one image generates a line l' in the other image on which its corresponding point x' must lie. From this fact it is clear that the search for correspondences is reduced from a 2D region to a line. This is illustrated in Figure 6.17.

A final detail is that the epipolar lines have a general orientation on the image plane. In order to simplify the computation, stereo algorithms

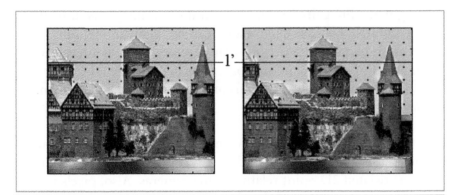

Figure 6.18: Rectified stereo image pair.

usually perform a *rectification* of the images. Rectification determines a transformation of each image plane such that pairs of conjugate epipolar lines become collinear and parallel to one of the image axes. The rectified images can be imagined as images acquired by a standard stereo setup obtained by rotating the original cameras. The main advantage of rectification is that computing stereo correspondences is reduced to a search along the horizontal lines of the rectified images. Figure 6.18 shows a pair of rectified images and one epipolar line.

Stereo correspondence with graph cuts

The dense stereo correspondence problem can be posed as follows. Every pixel $p(x, y)$ in the first image has a particular disparity d with respect to the matching pixel $q(x, y)$ in the second image. Then, we have set A of pixel pairs that may potentially correspond

$$A = \{(p, q) | p_y = q_y \text{ and } 0 \leq q_x - p_x \leq K\}, \qquad (6.16)$$

where $d = q_x - p_x$ is the disparity of the pair (p, q).

Our objective is to find a subset of A containing the actual corresponding pairs. This is equivalent to computing an assignment function $f(p, q)$ for $(p, q) \in A$, such that $f(p, q) = 1$ if pixels p and q correspond and $f(p, q) = 0$ otherwise. When the value of f is 1, the assignment is called *active*.

Here, it is assumed that the disparities d belong to the interval $[0, K]$. This is justified because in a stereo pair, the disparity tends to vary within a fixed range. Disparity also varies smoothly over the image, except at depth discontinuities. These characteristics make the dense stereo correspondence suitable to be posed as an energy-minimization problem that can be effectively solved by the graph cuts method (Section 6.6). Furthermore, energy models like the Potts energy can help incorporate the above contextual properties.

One of the aspects that make stereo correspondence a hard problem is the presence of *occlusions*. An occlusion occurs when a 3D point is visible in only one of the images of a stereo pair. Typically, occlusions will be present near depth discontinuities in the scene.

To model occlusions, we use the concept of *unique configuration*. Let $A(f)$ be the set of active assignments according to f, and $N_p(f) = \{(p, q) \in A(f)\}$ be the active assignments that involve the pixel p. A configuration f is called *unique* if each pixel is present in, at most, one active assignment. That is, $|N_p(f)| \leq 1$ for all pixels $p \in P$. Such condition guarantees the consistency of the correspondence. However, occluded pixels are identified by having $|N_p(f)| = 0$.

The energy function associated with a unique configuration f is composed of three terms:

$$E(f) = E_d(f) + E_o(f) + E_s(f). \qquad (6.17)$$

The first term is a data term that imposes a penalty based on the difference of intensities between corresponding pixels p and q

$$E_d = \sum_{(p,q) \in A(f)} D(p, q), \qquad (6.18)$$

where usually $D(p, q) = (I(p) - I(q))^2$.

The second term handles occlusions. It imposes a penalty for occluded pixels:

$$E_o = \sum_{p \in P} c_p T(|N_p(f)| = 0), \qquad (6.19)$$

where T is the indicator function and c_p is a constant.

The third term is the smoothness term that encourages configurations in which adjacent pixels have close disparity values.

$$E_s = \sum_{\{(p,q),(p',q')\} \in A} V((p,q),(p',q'))T(f(p,q) \neq f(p',q')), \quad (6.20)$$

where V is the Potts model for discontinuity preserving energy (see Eq. 6.5). Note that here the Potts energy is applied to assignments and not to pixels.

We remark that the energy function is finite only for unique configurations. A nonunique f has $E(f) = \infty$ and is not considered in the optimization.

Minimizing the correspondence energy function (6.17) is an NP-hard problem. But it can be solved approximately using the expansion-move algorithm of Section 6.6.4.

In order to apply the expansion-move algorithm, it is necessary to construct the graph G according to the structure of the problem and set the edge weights in G based on the energy function.

The key idea for extending the notion of α-expansions to the stereo problem is that the labels are disparities. Thus, the vertices of the graph correspond to assignments and the edges are links to assignments with the same label.

For an assignment $a = (p,q)$, let $d(a) = (p_x - q_x)$ be its disparity value. Let A^α be the set of all assignments in A for which $d(a) = \alpha$. In this setting, an α-expansion move on a configuration f is a subset of $A(f) \cup A^\alpha$, i.e., some assignments are replaced with new assignments with label (disparity) α.

The graph G will have active assignments as vertices. In an α-expansion, active assignments may become inactive and inactive assignments with disparity α may become active.

At each cycle, the graph is defined by the previous and current α assignments. Starting with a base configuration f^0, active assignments for a new configuration $\tilde{f}$ will be a subset $\tilde{A} = A^0 \cup A^\alpha$ such that $A^0 = \{a \in A(f^0) | d(a) \neq \alpha\}$ and $A^\alpha = \{a \in A | d(a) = \alpha\}$. The new configuration then is defined by $\tilde{f} A(\tilde{f}) = \tilde{A}$.

The vertices of G are given by the assignments $a \in \tilde{A}$ plus the terminals s, t. The edges in G fall into two categories: first, every a is connected to both s and t; second, for a pair of assignments (a_1, a_2) in the neighborhood system $N \subset \{(a_1, a_2) | a_1, a_2 \in A\}$ such that $a_1 = (p, q)$, $a_2 = (p', q')$ with (p, p') or (q, q') neighbor pixels, there will be edges (a_1, a_2) and (a_2, a_1), as well as vertices $a_1 = (p, q)$ and $a_2 = (p, r)$ with a common pixel p.

The edge weights follow in a straightforward manner from the energy function E and structure of the graph G.

Now, let us show an example of using graph cuts to solve the stereo correspondence problem. The example uses a standard data set from the University of Tsukuba. It is composed of a stereo image pair with hand-labelled integer disparities. Figure 6.19 depicts the data set. The images have a resolution of 384 by 288 pixels. In the example, the disparities have been quantized to 16 levels.

Figure 6.20 shows the result of using the α-expansion graph cuts algorithm with the Tsukuba stereo pair. Figure 6.20(a) shows the signed disparity error (where gray means zero). Figure 6.20(b) shows bad pixels (i.e., pixels with absolute disparity error greater than 1.0). Figure 6.20(c) shows computed disparities.

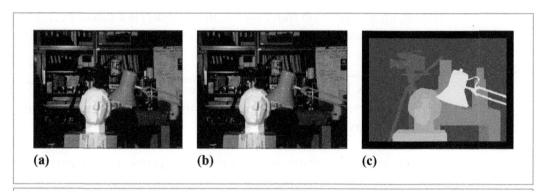

Figure 6.19: Tsukuba data set: (a) left image; (b) right image; (c) ground truth.

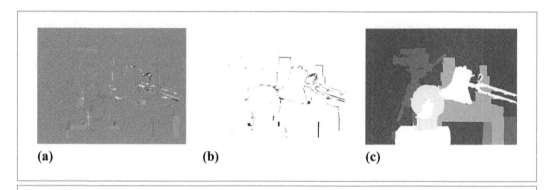

Figure 6.20: Graph cuts solution: (a) signed disparity error; (b) bad pixels; (c) disparities.

For comparison, in Figure 6.21(a), the solution was computed using the adaptive cost aggregation algorithm from Wang *et al.* (2006), and in Figure 6.21(b) the solution was computed using the region-tree algorithm of Lei *et al.* (2006). Observe that the results produced by these algorithms are clearly inferior to graph-cuts stereo.

Final Remarks

In this subsection, we discussed the problem of stereo matching using graph cuts. The stereo reconstruction is a specific case of a more

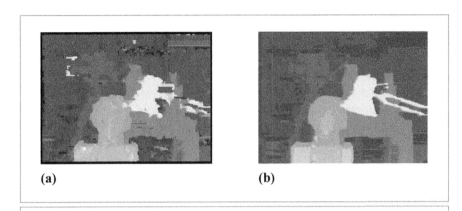

Figure 6.21: Stereo solution computed with other algorithms: (a) Wang *et al.* (2006) (b) Lei *et al.* (2006)

general problem of multicamera scene reconstruction, aiming to compute the 3D shape of an arbitrary scene from a set of *n* images taken from known viewpoints. This problem is considerably harder than stereo reconstruction esssentially due to the difficulty of reasoning about visibility. Nonetheless, the graph-cuts method has been applied successfully to multicamera reconstruction by Kolmogorov and Zabih (2002). The solution required the use of an energy function of class F^3.

6.9 COMMENTS AND REFERENCES

The general problem of multidimensional quantization is NP-complete. There exist algorithms based on several heuristics, such as "cluster analysis," that although do not guarantee an optimal solution, produce good results. The interested reader should take a look at Velho *et al.* (1997).

The problem of optimization for interactive visualization is closely related to the computation of meshes of variable resolution adapted to the view-dependent visualization conditions (Hoppe, 1997).

An evaluation of dense two-frame stereo correspondence algorithms was made by Scharstein and Szeliski (2002).

Graph-cuts methods have been used to solve many problems in computer graphics and vision. Some of the recent applications include interactive digital photomontage (Agarwala *et al.*, 2004); texture synthesis (Kwatra *et al.*, 2003), and foreground extraction (Marsh *et al.*, 2006; Rother *et al.*, 2004).

The shortest path algorithm has several applications in image processing and analysis. A popular technique for segmentation using shorthest path is called intelligent scissoring (Li *et al.*, 2004; Mortensen and Barrett, 1995, 1998; Wong *et al.*, 2000).

Discrete optimization can be used in computer animation to synthesize motion sequences from a database of motion clips. The database is typically built from motion-captured data. The basic idea is to generate complete sequences of motion by concatenating various short motion clips. In order to do that first the so-called *motion graph* is constructed. In this structure, every edge is a clip of motion, and nodes serve as choice points connecting clips. Any sequence of nodes, or walk, is itself a valid motion. Optimization techniques are used both to construct the motion graph and to extract motion sequences from a graph (Kovar *et al.*, 2002; Safonova and Hodgins, 2007).

The same idea of motion graphs can also be used in the context of video, in this case to generate cyclic video textures (Schödl *et al.*, 2000) or nonlinear editing structures (Sand and Teller, 2004).

BIBLIOGRAPHY

Agarwala, A., M. Dontcheva, M. Agrawala, S. Drucker, A. Colburn, B. Curless, D. Salesin, and M. Cohen. Interactive digital photomontage. *ACM Transactions on Graphics*, 23(3):294–302, 2004.

Boykov, Y., O. Veksler, and R. Zabih. Fast approximate energy minimisation via graph cuts. *IEEE Transactions on Pattern Analysis and Machine Intelligence*, 29:1222–1239, 2001.

Dahlhaus, E., D. S. Johnson, C. H. Papadimitriou, P. D. Seymour, and M. Yannakakis. The complexity of multiway cuts (extended abstract). In pages 241–251, *STOC'92: Proceedings of the Twenty-Fourth Annual ACM Symposium on Theory of Computing*. New York, NY: ACM Press, 1992.

Dinic, E. A. Algorithm for solution of a problem of maximum flow in a network with power estimation. *Soviet Mathematics*, 11:1277–1280, 1970.

Edmonds, J., and R. M. Karp. Theoretical improvements in algorithmic efficiency for network flow problems. *Journal of ACM*, 19(2):248–264, 1972.

Elias, P., A. Feinstein, and C. E. Shannon. Note on maximum flow through a network. *IRE Transactions on Information Theory*, 2:117–119, 1956.

Ford, L. R., and D. R. Fulkerson. *Flows in Networks*. Princeton, NJ: Princeton University Press, 1969.

Fuchs, H., Z. Kedmen, and S. Uselton. Optimal surface reconstruction from planar contours. *Communications of the ACM*, 20(10):693–702, 1977.

Funkhouser, T. A., and C. H. Séquin. Adaptive display algorithm for interactive frame rates during visualization of complex virtual environment. In T. Kajiya (Ed.), *Computer Graphics (SIGGRAPH' 93 Proceedings)*. New York, NY: Vol. 27, 247–254, 1993.

Gale, D. H. W. Kuhn, and A. W. Tucker. Linear programming and the theory of games. *Activity Analysis of Production and Allocation*, 317–329, 1951.

Goldberg, A. V., and R. E. Tarjan. A new approach to the maximum-flow problem. *Journal ACM*, 35(4):921–940, 1988.

Hoppe, H. View-dependent refinement of progressive meshes. In *Proceedings of SIGGRAPH 97. Computer Graphics Proceedings,*

Annual Conference Series. SIGGRAPH, New York, NY: 189–198, 1997.

Kolmogorov, V. *http://www.cs.cornell.edu/rdz/graphcuts.html*, 2004.

Kolmogorov, V., and R. Zabih. Multi-camera scene reconstruction via graph cuts. In *ECCV '02: Proceedings of the 7th European Conference on Computer Vision-Part III.* London, UK: Springer-Verlag, 82–96, 2002.

Kolmogorov, V., and R. Zabih. What energy functions can be minimized via graph cuts? *IEEE Transactions on Pattern Analysis and Machine Intelligence,* 26(2):147–159, 2004.

Kovar, L., M. Gleicher, and F. Pighin. Motion graphs. *ACM Transactions on Graphics,* 21(3):473–482, 2002.

Kwatra, V., A. Schödl, I. Essa, G. Turk, and A. Bobick. GraphCut textures: Image and video synthesis using graph cuts. *ACM Transactions on Graphics.* 22(3):277–286, 2008.

Lei, C., J. Selzer, and Y. H. Yang. Region-tree based stereo using dynamic programming optimization. In *Proceedings of CVPR - IEEE Computer Society Conference on Computer Vision and Pattern Recognition,* Washington, DC, 23378–2385, 2006.

Li, S. Z. *Markov Random Field Modeling in Computer Vision.* London, UK: Springer-Verlag, 1995.

Li, Y., J. Sun, C.-K. Tang, and H.-Y. Shum. Lazy snapping. *ACM Transactions on Graphics,* 23(3):303–308, 2004.

Marsh, M., S. Bangay, and A. Lobb. Implementing the "GrabCut" segmentation technique as a plugin for the GIMP. *AFRIGRAPH 2006,* 171–176, 2006.

Mason, A. E. W., and E. H. Blake. Automatic hierarchical level of detail optimization in computer animation. *Computer Graphics Forum,* 16(3):191–200, 1997.

Meyers, D., S. Skinner, and K. Sloan. Surfaces from contours: The correspondence and branching problems. In *Proceedings of Graphics Interface '91.* San Diego, CA, 246–254, 1991.

Mortensen, E. N., and W. A. Barrett. Intelligent scissors for image composition. In *Proceedings of SIGGRAPH 95. Computer Graphics*

Proceedings, Annual Conference Series. New York, NY: 191–198, 1995.

Mortensen, E. N., and Barrett, W. A. Interactive segmentation with intelligent scissors. *Graphical Models and Image Processing,* 60(5):349–384, 1998.

Rother, C., V. Kolmogorov, and A. Blake. GrabCut: Interactive foreground extraction using iterated graph cuts. *ACM Transactions on Graphics,* 23(3): 309–314, 2004.

Safonova, A., and J. K. Hodgins. Construction and optimal search of interpolated motion graphs. *ACM Transactions on Graphics,* 26(3):106, 2007.

Sand, P., and S. Teller. Video matching. *ACM Transactions on Graphics,* 23(3):592–599, 2004.

Scharstein, D., and R. Szeliski. A taxonomy and evaluation of dense two-frame stereo correspondence algorithms. *International Journal of Computational Vision,* 47(1-3):7–42, 2002.

Schödl, A., R. Szeliski, D. H. Salesin, and I. Essa. Video textures. *ACM Press Association for Computing Machinery. Computer Graphics Proceedings, Annual Conference Series.* New York, NY: ACM Press, 489–498, 2000.

Velho, L., J. Gomes, and M. R. Sobreiro. Color image quantization by pairwise clustering. In *SIBGRAPI X, Brazilian Symposium Of Computer Graphics and Image Irocessing.* Campos do Jordao, SP, Los Alamitos, CA: IEEE Computer Society, 203–210, 1997.

Wang, L., M. Liao, M. Gong, R. Yang, and D. Nister. High-quality real-time stereo using adaptive cost aggregation and dynamic programming. *3DPVT '06: Proceedings of the Third International Symposium on 3D Data Processing, Visualization, and Transmission (3DPVT'06).* Washington, DC: IEEE Computer Society, 798–805, 2006.

Wong, K. C.-H., P.-A. Heng, and T.-T. Wong. Accelerating "intelligent scissors" using slimmed graphs. *Journal of Graphics Tools,* 5(2):1–13, 2000.

Wu, X. Color quantization by dynamic programming and principal analysis. *ACM Transactions on Graphics*, 11(2):348–372, 1992.

Zhou, Y., A. Kaufman, and A. W. Toga. 3D skeleton and centerline generation based on an approximate minimum distance field. *The Visual Computer*, 14(7):303–314, 1998.

7 GLOBAL OPTIMIZATION

In the preceeding chapters, we have seen general-purpose methods for continuous and discrete problems. The methods for continuous problems were only able to find local minima, in general. In the case of discrete problems, the solution space is frequently too large to find the global minima in a reasonable time; we need to use heuristics. In this chapter, we describe some approaches to finding the global minima, using heuristics for discrete problems and guaranteed methods for continuous problems.

7.1 TWO EXTREMES

As we have seen in Chapter 2, many optimization problems can be formulated in the following way: among a finite set of candidates, find one that best fits a given criterion. Since the set of candidates is finite, such problems are apparently easy to solve, using *exhaustive enumeration*. We simply enumerate *all* candidates in succession, evaluate the optimality criterion on each candidate, and select the best candidate at the end. In practice, this solution does not work because the set of candidates is too large although it is finite. This happens for all the main combinatorial problems, such as the famous traveling salesman problem.

For problems whose candidate space is huge, an alternative approach to their solution is to use stochastic methods. We choose a candidate at random and evaluate the optimality criterion on it. We repeat this process many times, and we select the best candidate found. In other words, stochastic methods perform a "random walk" in the space of candidates.

These two approaches—exhaustive enumeration and random walk— are two extreme examples of methods that explore the space of candidates. Exhaustive enumeration is systematic and explores the candidate space completely, which can be very slow for huge spaces. Random walk is not at all systematic and explores the candidate space in no particular order and without any method. In both cases, the only information "learned" by the method is which is the best candidate at a given moment. None of the methods learns or exploits the *structure* of the space of candidates.

7.2 LOCAL SEARCH

A more clever solution is to use *local search*, also known as "hill climbing." Starting at an initial candidate, we "move" always in the most promising direction, i.e., the direction in which the objective function grows (Figure 7.1). When the objective function has a single global maximum, then this solution can be very efficient, but in general, this method tends to get caught and stuck in local maxima because, at these points, the objective function does not grow in *any* direction (i.e., there is no "hill" to climb). In Figure 7.1, starting the "climb" at the black dot, we can reach the gray dot, which is a local maximum, but we cannot reach the higher peaks on the right and on the left.

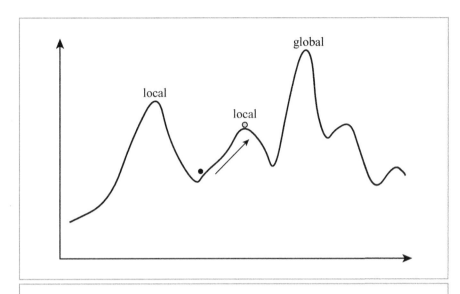

Figure 7.1: Local maxima × global maxima in hill climbing (Beasley *et al.*, 1993).

7.3 STOCHASTIC RELAXATION

From the previous discussion, it is clear that efficient solutions to the global optimization problem must use methods that perform a systematic—but not exhaustive—exploration of the space of candidates and that are still flexible enough to try several different candidates to avoid getting stuck in local maxima.

The technique of *stochastic relaxation*, also known as *simulated annealing*, can be seen as a clever combination of local search and random walk. We start at a random point in the space of candidates. At each step, we move in a random direction. If this movement takes us to a higher point, then we accept this point as the new current point. If this movement takes us to a lower point, then we accept this point with probability $p(t)$, where t is the simulation time. The value of the function p starts near 1 but decreases to 0 when t grows, typically in an exponential way. Thus, our exploration of the space of candidates starts as a random walk and ends as a local search.

The physical analogy that motivates this method is the process of cooling of a solid object: when the temperature is high, the solid is soft and easily shaped; as the temperature increases, the solid becomes more rigid, and it is more difficult to change its shape.

Simulated annealing is able to escape from local minima, but because only one candidate is considered in turn, this method also learns very little about the structure of the space of candidates.

7.4 GENETIC ALGORITHMS

Another class of flexible systematic methods for global optimization is the class of *genetic algorithms*. Algorithms in this class are based on ideas from biology, which we now review briefly.

7.4.1 GENETIC AND EVOLUTION

In the optimization problems that we shall consider, the space of candidates can be parametrized by a small number of parameters, as described in Section 2.1. Using the terminology of biology, each candidate represents an *individual* of a *population*. Each individual is determined by the *genetic content*, or *genotype*, which is stored in *chromosomes*.[1] Each parameter is called a *gene*. Each gene controls one aspect of the individual and can take only a finite number of different states (or values), called *alleles*. The set of gene values completely determines the individual; we say that the individual is the *phenotype* or expression of the genotype.

---

1   In nature, individuals possess many chromosomes; for instance, human beings possess 46 chromosomes, totaling fifty to hundred thousand genes.

Individuals are immersed in an environment, which is under continuous change. According to Darwin,[2] individuals in nature compete among themselves to get food, shelter, partners, etc.; only the individuals who are able to face environmental changes manage to survive, reproduce, and preserve their characteristics. In other words, the *evolution* of the individuals is performed by the *selection* of the fittest. Genetic algorithms use the objective function (or a monotone variant of it) as a fitness criterion of individuals.

Two genetic mechanisms are essential so that a population can evolve and adapt itself to changes in the environment: mutations and sexual reproduction. *Mutations* are random variations in genotypes. Some mutations are favorable to survival; other mutations are unfavorable, and the individuals who carry them are eliminated by selection. Thus, the existence of mutations allows populations to adapt to environmental changes. However, if mutations were the sole adaptation mechanism, evolution would be extremely slow because nature is conservative: mutations are not very frequent, and they are not always favorable. Moreover, an individual would never have only favorable mutations. *Sexual reproduction*, which randomly mixes the genetic material of one or more individuals, allows a much faster evolution because it allows individuals to inherit favorable characteristics from more than one source.

7.4.2 GENETIC ALGORITHMS FOR OPTIMIZATION

Inspired by the theory of evolution described in the previous section, *genetic algorithms* explore the candidate space of an optimization problem in a way that is both systematic and stochastic, as described below:

1. Begin with an initial population with random genotypes or computed using some heuristic that is specific to the given problem. Each gene represents a parameter of the candidate space.

---

2   Charles Darwin (1809–1882), English naturalist, who founded the modern theory of evolution by natural selection, described in his classic work *The Origin of Species.*

2. At each step of the evolution, some individuals have mutations and some individuals reproduce.

3. All individuals are evaluated according to a fitness criterion that is based on the objective function. Only the fittest individuals survive to the next generation.

4. Evolution stops when the fittest of the best individual has converged to some value (that we hope is close to the global optimum) or after a fixed number of steps.

There are many variations of this basic scheme; for instance, steps 2 and 3 can be interchanged: only the fittest individuals have mutations or reproduce.

There are many details to be settled and many parameters to be adjusted: the representation of genotypes; the size and nature of the initial population; the frequency and nature of mutations and reproductions; the fitness criterion for choosing the best individuals; the criteria for convergence and divergence of the evolution. The art in the use of genetic algorithms lies exactly in handling these details according to each problem to be solved.

In the original work of Holland (1975), who introduced the theoretical basis for genetic algorithms, genotypes contained only one chromosome, which was represented by bit strings, one bit for each gene. In this case, the basic genetic operations—mutation and reproduction—are easy to define and understand. A mutation in a bit string is simply a change in a single bit: from 0 to 1 or from 1 to 0. Sexual reproduction mimics the crossover mechanism that occurs in nature: the chromosomes of the parents are cut at a crossing point and recombined in a crossed fashion, yielding two offsprings, as shown in Figure 7.2. These are the simplest rules; more complicated rules can (and should) be formulated for specific situations.

More recently, Koza (1992) proposed the use of *programs* instead of bit chains as genotypes in the exploration of spaces that are not naturally described by a finite number of parameters but in which the candidates still have a finite description. Sims (1991a) was the first to use

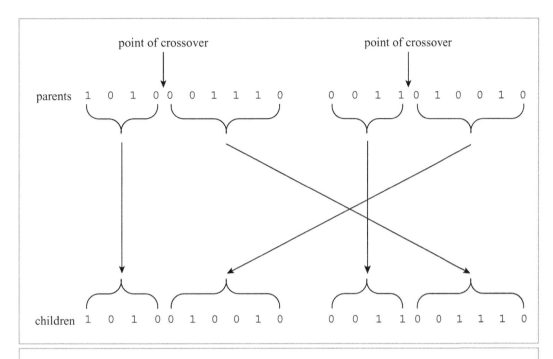

Figure 7.2: Sexual reproduction by chromosomal crossover (Beasley *et al.*, 1993).

this kind of genetic algorithm in computer graphics for image synthesis and modeling.

7.4.3 THE TRAVELING SALESMAN PROBLEM

As a concrete example of the use of genetic algorithms in discrete optimization problems, consider the classic traveling salesman problem. We wish to visit n cities in a region using a closed circuit of least length. If we number the cities from 1 to n, a tour for the traveling salesman can be represented by a permutation of the numbers from 1 to n.

Michalewicz (1994) describes the following genetic model for this problem. Individuals represent the possible tours. The fitness of an individual is given by the length of the tour that the individual represents, which can be computed by adding the distance between consecutive cities in the tour. The shortest the tour, the fitter the individual.

Each individual has a single chromosome with n genes; each gene can take integer values from 1 to n. (Note that genes are not bit strings in this formulation.) Not all possible chromosomes represent valid individuals because only permutations are valid. Thus, the candidate space has $n!$ elements. (For $n = 100$, we have $100! \approx 10^{158}$ elements!) Mutation occurs by changing the value of one or more genes, or the order of the cities in a chromosome. Crossed reproduction combines segments of the two tours. If these operations do not take the nature of the problem into account, it is possible (and even probable) to generate chromosomes that do not represent permutations. These chromosomes must be considered defective, and the individuals do not survive. In this case, the candidate space actually has n^n elements, only $n!$ of which are considered valid.

Michalewicz (1994) reports that, for $n = 100$, starting with an initial random population, after 20000 generations, we typically obtain a solution to the traveling salesman problem whose length is less than 10% above the optimal value.

7.4.4 OPTIMIZATION OF A UNIVARIATE FUNCTION

We shall now see an example of continuous optimization (also from Michalewicz, 1994). We want to maximize the function $f(x) = x \sin(10\pi x) + 1$ in the interval $[-1, 2]$. As shown in Figure 7.3, this function has many local maxima but only one global maximum, which occurs at $x \approx 1.85$.

Michalewicz (1994) describes the following genetic model for this problem. Individuals represent real numbers, and their fitness is measured directly by the function f: the larger the value of f, the fitter the individual.

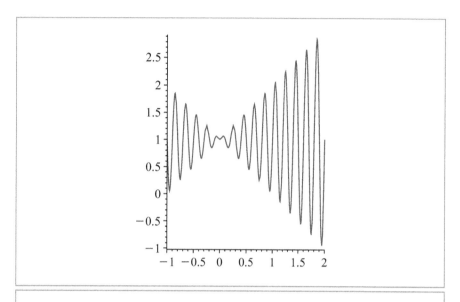

Figure 7.3: Graph of $f(x) = x\sin(10\pi x) + 1$ on $[-1, 2]$.

Each individual has a single chromosome with n genes (each gene is a single bit). The individual corresponding to one such chromosome is a point in the uniform mesh of points that has 2^n points in the interval $[-1, 2]$. More precisely, if the genotype is

$$x = (x_n\ldots x_1),$$

where each x_i is a bit, the the phenotype is the following real number:

$$x = -1 + \frac{3}{2^n - 1} \sum_{i=1}^{n} x_i 2^{i-1}.$$

In other words, the integers from 0 to $2^n - 1$, which are exactly those representable with n bits, are linearly mapped to the interval $[-1, 2]$. This mapping transforms the continuous problem into a discrete problem, whose solution approximates the solution of the continuous problem. This approximation can be as good as desired: we just need to use a sufficiently fine discretization, i.e., a sufficiently large number of bits in the chromosomes.

Mutations occur by randomly changing one or more bits, with probability given by a user-defined mutation rate. To mutate a bit in a genotype x, we randomly select a position k between 1 and n and we set $x_k \leftarrow 1 - x_k$. Two individuals mate by mixing their bits: we randomly select a position k between 0 and n; the chromosomes of the parents are split after position k and recombined, yielding two offsprings. More precisely, if the parents are $a = (a_n \ldots a_1)$ and $b = (b_n \ldots b_1)$, then the offsprings are $(a_n \ldots a_k b_{k-1} \ldots b_1)$ and $(b_n \ldots b_k a_{k-1} \ldots a_1)$ (see Figure 7.2).

Michalewicz (1994) reports a solution to this problem that locates the global maximum with a precision of six decimal places. For this, he used $n = 22$, an initial random population of 50 individuals, mutation rate equal to 0.01, and crossing rate equal to 0.25. After 150 generations, the fittest individual had genotype 1111001101000100000101. This individual corresponds to the approximate solution $\hat{x} = 2587568/1398101 \approx 1.850773$, whose fitness is $f(\hat{x}) \approx 2.850227$. The global maximum of f is $f^* \approx 2.850273$, attained at $x^* \approx 1.850547$. Thus, the result has an absolute error less than 5×10^{-5} and a relative error less than 2×10^{-5}—a good solution.

7.5 GUARANTEED GLOBAL OPTIMIZATION

In this section, we revisit the unconstrained continuous minimization problem, seen in Chapter 4. Given a box[3] $\Omega \subseteq \mathbb{R}^d$ and an objective function $f : \Omega \to \mathbb{R}$, find its *global minimum* $f^* = \min\{ f(x) : x \in \Omega \}$ and the set of all points in Ω where this minimum value is attained, $\Omega^*(f) = \{ x^* \in \Omega : f(x^*) = f^* \}$.

---

3   A *box* is the Cartesian product of d real intervals: $[a_1, b_1] \times \cdots \times [a_d, b_d]$.

In Chapter 4, we saw methods that find *local* minima. In this section, we are interested in finding the *global* minimum. In general, it is not possible to find this minimum exactly, and so we shall consider an approximate, numerical version of the problem. Instead of finding f^* and all minimum points $\Omega^*(f)$ exactly, we seek only to identify some real interval M that is *guaranteed* to contain the global minimum f^* and some subset $\hat{\Omega}$ of Ω that is *guaranteed* to contain $\Omega^*(f)$. The goal then is to make M and $\hat{\Omega}$ as small as possible for a given computation budget (i.e., time or memory).

Methods that sample the objective function f at a finite set of points in Ω cannot reliably find the global minimum f^* because f may oscillate arbitrarily between these sample points.

Consider the following example (Moore, 1991). Choose a small subinterval $[x_1, x_2] \subseteq \Omega$ and define

$$g(x) = f(x) + \begin{cases} 0, & \text{for } x \notin [x_1, x_2] \\ (x - x_1)(x - x_2)10^{20}, & \text{for } x \in [x_1, x_2]. \end{cases}$$

The function g coincides with f, except on the interval $[x_1, x_2]$ in which it attains a very small minimum. If this interval is sufficiently small (for instance, if x_1 and x_2 are consecutive floating-point numbers), then it is not possible to notice the difference between f and g by point sampling.

What is needed for the reliable solution to the global optimization problems is to replace point sampling by reliable estimates of the *whole* set of values taken by f on a subregion X of Ω. (This would allow us to notice the difference between f and g in the previous example.) More precisely, if for each $X \subseteq \Omega$ we know how to compute an interval $F(X)$ that is *guaranteed* to contain $f(X) = \{f(x) : x \in X\}$, then it is possible to eliminate parts of Ω that cannot contain global minimizers. The parts that cannot be eliminated are subdivided so that after reaching a user-defined tolerance, whatever is left[4] of Ω must contain *all* global minimizers and so serves as an approximate solution $\hat{\Omega}$. The aim of this section is precisely how to compute such *interval estimates* $F(X)$ and how to use those estimates to eliminate parts of Ω.

7.5.1 BRANCH-AND-BOUND METHODS

A branch-and-bound algorithm for unconstrained global minimization alternates between two main steps: *branching*, which is a recursive subdivision of the domain Ω, and *bounding*, which is the computation of lower and upper bounds for the values taken by f in a subregion of Ω. By keeping track of the current best upper bound $\hat{f}$ for the global minimum f^*, one can discard all subregions whose lower bound for f is greater than $\hat{f}$ because they cannot contain a global minimizer of f. Subregions that cannot be thus discarded are split and the pieces are put into a list $\mathcal{L}$ to be further processed. Thus, at all times, the set $\hat{\Omega} = \cup \mathcal{L}$, given by the union of all subregions in $\mathcal{L}$, contains *all* possible global minimizers and so is a valid solution to the global minimization problem, as defined at the beginning of this section.

More precisely, the branch-and-bound solution to the global minimization problems starts with $\mathcal{L} = \{\Omega\}$ and $\hat{f} = \infty$ and repeats the steps below while $\mathcal{L}$ is not empty:

1. select one subregion from $\mathcal{L}$
2. if X is small enough, then accept X as part of $\hat{\Omega}$
3. compute an interval estimate $F(X)$ for $f(X)$
4. if $\inf F(X) > \hat{f}$, then discard X
5. update $\hat{f} \leftarrow \min(\hat{f}, \sup F(X))$
6. subdivide X into X_1 and X_2
7. add X_1 and X_2 to $\mathcal{L}$.

---

4   "When you have eliminated the impossible, whatever remains, however improbable, must be the truth."—Sherlock Holmes in *The Sign of Four*, Arthur Conan Doyle (1889).

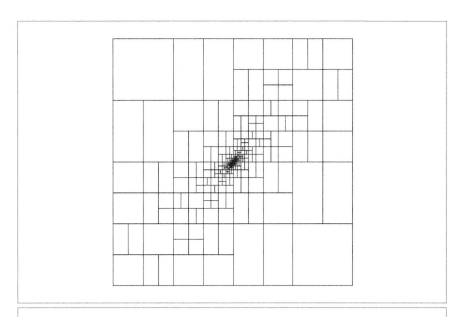

Figure 7.4: Domain decomposition performed by branch-and-bound.

Figure 7.4 shows the decomposition of $\Omega = [-10, 10] \times [-10, 10]$ obtained by a branch-and-bound algorithm to minimize Matyas's function

$$f(x_1, x_2) = 0.26\,(x_1^2 + x_2^2) - 0.48\,x_1 x_2,$$

whose global minimum $f^* = 0$ is attained at $\Omega^* = \{(0, 0)\}$. The white regions have been discarded, and the approximate solution $\hat{\Omega}$ appears in gray.

The basic branch-and-bound algorithm outlined above admits endless variations, depending on how the selection, branching, and bounding steps are implemented.

The main criteria for selecting from $\mathcal{L}$ the next box X to be examined at each step are as follows:

- X is the newest box in $\mathcal{L}$. In this case, the list $\mathcal{L}$ is being used as a stack and Ω is explored in depth.

- X is the oldest box in $\mathcal{L}$. In this case, the list $\mathcal{L}$ is being used as a queue and Ω is explored in breadth.
- X is the box in $\mathcal{L}$ that has the smallest lower bound for f. In this case, the list $\mathcal{L}$ is being used as a priority queue and Ω is explored at each step by giving priority to the regions that seem more promising to contain a global minimizer.

The simplest branching method is to bisect the current box orthogonally to its widest direction. Cyclic bisection (Moore, 1979) is another natural choice for branching. For reviews of branching methods and strategies for selecting subregions from $\mathcal{L}$, including recent results, see Berner (1996), and Csendes and Ratz (1997).

In Section 7.6, we describe how to implement the bounding step, i.e., how to compute interval estimates $F(X)$ for $f(X)$.

7.5.2 ACCELERATING INTERVAL BRANCH-AND-BOUND METHODS

Due to overestimation in interval estimates $F(X)$, pure branch-and-bound algorithms tend to spend a great deal of time trying to eliminate regions close to local minimizers. Therefore, in practice, it is essential to use additional tests to accelerate convergence.

The simplest additional test is the *midpoint test*. At each step, the objective function f is evaluated at the center of the current subregion X, and this value is used to update the current best upper bound $\hat{f}$ for the global minimum f^*. This test tends to compensate for possible overestimation in $F(X)$ by using an actual point of $f(X)$.

The following conditions are sufficient for the convergence of interval branch-and-bound algorithms: the objective function f is continuous, the branching step reduces the diameter of boxes to zero, and the diameter of $F(X)$ tends to zero as the diameter of X goes to zero. Better results are possible when f is sufficiently differentiable; we can apply the necessary first- and second-order tests discussed in Chapter 4.

The *monotonicity test* computes interval estimates for the components of the gradient ∇f in a subregion X. If any of these estimates does not contain zero, then f is monotone in the corresponding direction and X cannot contain a local minimizer. The *convexity test* computes interval estimates for the components of the Hessian $\nabla^2 f$ in X. If these estimates show that the Hessian matrix is positive definite over X, then again X cannot contain a local minimizer.

These high-order tests are more expensive but can discard large subregions of Ω. Moreover, they can be combined with the methods discussed in Chapter 4 to locate subregions X that contain a single local minimizer, which is then computed with a local method (not an interval method), thus avoiding excessive subdivision of X.

7.5.3 CONSTRAINED OPTIMIZATION

The branch-and-bound method described above for unconstrained problems can be adapted for solving constrained problems:

$$
\begin{aligned}
\min \quad & f(x) \\
\text{subject to} \quad & g_i(x) = 0, \quad i = 1, \ldots, m \\
& h_j(x) \geq 0, \quad j = 1, \ldots, p.
\end{aligned}
$$

In addition to the interval estimate F for f, we also need interval estimates G_i and H_j for the restrictions g_i and h_j, respectively. At each step, we test whether the subregion X under examination is feasible and discard it if it is not. More precisely, if $G_i(X)$ does not contain zero, then the restriction $g_i(x) = 0$ is not satisfied in X; if $H_j(X)$ contains only negative numbers, then the restriction $h_j(x) \geq 0$ is not satisfied in X. Subregions that are too small and have not been eliminated are considered feasible. Details of this method can be seen in Baker Kearfott (1996), Hansen (1988), and Ratschek and Rokne (1988).

An important subproblem is to determine whether a constrained problem is feasible, i.e., whether it is possible to satisfy all given restrictions in Ω. Many times, the goal is precisely to find the feasible region:

$$\widetilde{\Omega} = \{\, x \in \Omega : g_i(x) = 0, h_j(x) \geq 0 \,\}.$$

In these cases, the objective function is not relevant and can be taken constant. The goal then becomes simply to solve the system of equations and inequalities given by the restrictions. In this case, the elimination of those parts that are not feasible, as described above, is sufficient to yield an approximation $\hat{\Omega}$ for $\widetilde{\Omega}$.

7.6 RANGE ANALYSIS

Range analysis is the study of the behavior of real functions based on interval estimates for their set of values. More precisely, given a function $f : \Omega \to \mathbb{R}$, range analysis provides an *inclusion function* for f, i.e., a function F defined on the subsets of Ω such that $F(X)$ is an interval containing $f(X)$, for all $X \subseteq \Omega$. Thus, the inclusion function F provides robust estimates for the maximum and minimum values of f in X. The estimates provided by F must be robust but are not required to be tight. In other words, $F(X)$ has to contain $f(X)$ but may be strictly larger. Robust interval estimates, hence, play an essential role in the correctness of interval branch-and-bound algorithms.

Any method that uses an inclusion function to solve a global optimization problem actually provides a *computational proof* that the global minimum is in the interval M and that *all* points of Ω where this minimum is attained are in $\hat{\Omega}$, which no method based on point sampling can do.

The efficiency of interval methods depends on the quality of the inclusion functions used, i.e., on how close the estimate $F(X)$ is to the exact range $f(X)$. Here, as elsewhere, one usually trades quality for speed: techniques that provide tighter bounds are usually more expensive. However, tighter estimates often allow the algorithm to discard many regions that contain no minimizer at an earlier stage before they get subdivided, indicating that less range estimations need to be performed, and the algorithm runs faster as a whole.

There are special range analysis techniques for some classes of functions. If f is a Lipschitz function with Lipschitz constant L, then by definition we have

$$|f(x) - f(y)| \le L|x - y|,$$

for all $x, y \in X$. So, a simple and fast estimate for $f(X)$ is given by

$$F(X) = [f(x_0) - L\delta, f(x_0) + L\delta],$$

where x_0 is an arbitrary point of X and $\delta = \max\{|x - x_0| : x \in X\}$. For instance, we can take x_0 as the midpoint of X and δ as half the diameter of X. Note that if we know how to estimate the derivative of f, then the mean value theorem implies that f is a Lipschitz function.

If f is a polynomial function of degree n in the interval $[0, 1]$, then we can represent f in the Bernstein–Bézier basis of degree n:

$$f(x) = \sum_{i=0}^{n} b_i B_i^n(x),$$

where

$$B_i^n(x) = \binom{n}{i} x^i (1 - x)^{n-i}.$$

From this representation, we can take $F(X) = [\min b_i, \max b_i]$ because

$$B_i^n(x) \ge 0 \quad \text{and} \quad \sum_{i=0}^{n} B_i^n(x) = 1.$$

A similar result holds for multivariate function in domains of the form $\Omega = [a_1, b_1] \times \cdots \times [a_d, b_d]$.

We shall now describe interval arithmetic, a classical range analysis of wide applicability.

7.6.1 INTERVAL ARITHMETIC

One of the main problems with the theory and practice of numerical algorithms is the control of errors due to representation of the real numbers as the discrete set of floating-point numbers of fixed precision in use in digital computers. These *rouding errors* are not the same as the *approximation errors* in each algorithm, which occur because the desired solutions cannot be found using a finite number of elementary operations. Rounding errors are present even in elementary arithmetic operations, such as addition and multiplication.

A simple and a common approach to estimating the error in a floating-point computation is to repeat it using more precision and compare the results. If the results agree in many decimal places, then the computation is assumed to be correct, at least in the common part. However, this common practice can be seriously misleading, as shown by the following simple example (Rump, 1988).

Consider the evaluation of

$$f = 333.75y^6 + x^2(11x^2y^2 - y^6 - 121y^4 - 2) + 5.5y^8 + x/(2y),$$

for $x = 77617$ and $y = 33096$. Note that x, y, and all coefficients in f are exactly representable in floating point. Thus, the only errors that can occur in the evaluation of f are rounding errors. Rump (1988) reports that computing f in FORTRAN on an IBM S/370 mainframe yields

$f = 1.172603\ldots$ using single precision
$f = 1.1726039400531\ldots$ using double precision
$f = 1.172603940053178\ldots$ using extended precision.

Since these three values agree in the first seven places, common practice would accept the computation as correct. However, the true value is $f = -0.8273960599\ldots$; not even the *sign* is right in the computed results! (The problem does not occur only in FORTRAN or only in the IBM S/370. Repeat this computation using your favorite programming language or calculator.)

Interval arithmetic (IA) was invented by Moore (1966) with the explicit goal of improving the reliability of numerical computation. Even if it cannot make rounding errors disappear, it can at least be used to track the occurrence of damaging rounding errors and provide a measure of the impact of rounding errors in the final result. For instance, in Rump's example, IA would probably return a very large interval, which still contained the correct answer but too large to give its exact location. Nevertheless, a very large interval would certainly alert the user that the computation was severely damaged by rounding errors.

For our purposes, the most important feature of IA is that it is the natural tool for range analysis (Moore, 1966, 1979; Ratschek and Rokne, 1984), allowing us to implement reliable global optimization methods (Baker Kearfott, 1996; Hansen, 1988; Ratschek and Rokne, 1988).

In IA, a real number x is represented by a pair of floating-point numbers $[a, b]$, corresponding to an interval that is guaranteed to contain x, such that $a \leq x \leq b$. In this way, IA provides not only an estimate for the value of x (the midpoint $(a + b)/2$) but also bounds on how good this estimate is (the width of the interval, $b - a$).

The real power of IA is that we can operate with intervals as if they were numbers and obtain reliable estimates for the results of numerical computations, even when implemented in floating-point arithmetic.

To implement a numerical algorithm with IA, it is enough to use interval versions of the elementary operations and function composition. For instance, we have the following interval formulas for the elementary arithmetic operations:

$$[a, b] + [c, d] = [a + c, b + d]$$
$$[a, b] - [c, d] = [a - d, b - c]$$
$$[a, b] \times [c, d] = [\min(ac, ad, bc, bd), \max(ac, ad, bc, bd)]$$
$$[a, b]/[c, d] = [a, b] \times [1/d, 1/c], \quad \text{if } 0 \notin [c, d].$$

With a little extra work, we can write formulas for all elementary functions, such as square root, sine, cosine, logarithm, and exponential.

Therefore, if we know how to write an algorithm for computing a function f on n variables $x_1, \ldots, x_n$, then we know how to write an algorithm for computing an interval estimate $F(X)$ for f over $X = [a_1, b_1] \times \cdots \times [a_n, b_n]$: simply compose the interval versions of the elementary operations and functions in the same way they are composed to evaluate f. This is specially elegant to implement with programming languages that support operator overloading, such as C++, but IA can be easily implemented in any programming language. There are several implementations available on the Internet (Kreinovich, n.d.)

The Dependency Problem

A limitation of IA is that its range estimates tend to be too conservative: the interval $F(X)$ computed as estimate for the range of values of an expression f over a region X can be much wider than the exact range $f(X)$. This overconservatism of IA is particularly severe in long computation chains—such as the ones in computer graphics—and one frequently gets interval estimates that are too large, sometimes to the point of uselessness. This happens even in computations that are not severely affected by rounding errors.

The main reason for this overconservatism is the implicit assumption that operands in primitive interval operations are mutually independent. If there are dependencies among the operands, then those formulas are not the best possible because not all combinations of values in the operand intervals will be attained; the result interval will then be much wider than the exact range of the result quantity. This problem is called the *dependency problem* in IA.

An extreme example of this problem is the expression $x - x$. Although its value is always zero, this expression is computed blindly in IA as

$$x = [a, b]$$
$$x - x = [a - b, b - a].$$

Thus, the width of the interval $x - x$, which should be zero, is twice the width of x.

A less extreme and a more typical example is $x(10 - x)$ for x in $[4, 6]$:

$$x = [4, 6]$$
$$10 - x = [10, 10] - [4, 6] = [4, 6]$$
$$x(10 - x) = [4, 6] \cdot [4, 6] = [16, 36].$$

Note that the result $[16, 36]$ is 20 times larger than $[24, 25]$, which is the exact range of $x(10 - x)$ over $[4, 6]$. This discrepancy is due to the inverse dependency that exists between x and $10 - x$, a dependency that is not taken into account by the interval multiplication formula. For instance, the upper bound 36 comes from the product of the two upper bounds, which in this case are both equal to 6. However, these two upper bounds are never attained simultaneaously: when $x = 6$, we have $10 - x = 4$, and vice versa.

7.6.2 AFFINE ARITHMETIC

Affine arithmetic (AA) is a model for numeric computation designed to handle the dependecy problem (Comba and Stolfi, 1993). Like standard IA, AA can provide guaranteed bounds for the computed results, taking into account input, truncation, and rounding errors. Unlike IA, AA automatically keeps track of *correlations* between computed and input quantities and is therefore frequently able to compute much better estimates than IA, specially in long computation chains.

In AA, each quantity x is represented by an *affine form*

$$\hat{x} = x_0 + x_1\varepsilon_1 + x_2\varepsilon_2 + \cdots + x_n\varepsilon_n,$$

which is a polynomial of degree 1 with real coefficients x_i in the symbolic variables ε_i (called *noise symbols*) whose values are unknown but assumed to lie in the interval $[-1, +1]$. Each noise symbol ε_i stands for an independent component of the total uncertainty of the quantity x; the corresponding coefficient x_i gives the magnitude of that component.

The main benefit of using affine forms instead of intervals is that two or more affine forms share common noise symbols, indicating that the quantities they represent are at least partially dependent on each other. Since this dependency is represented in AA, albeit implicitly, AA can provide better interval estimates for complicated expressions, as the simple example in the end of this section shows.

To obtain interval estimates with AA, first, convert all input intervals to affine forms. Then, operate on these affine forms to compute the desired function by replacing each primitive operation with its AA version. Finally, convert the resulting affine form back into an interval. We shall now see these three steps in some detail.

Given an interval $[a, b]$ representing some quantity x, an equivalent affine form is $\hat{x} = x_0 + x_k \varepsilon_k$, where $x_0 = (b + a)/2$ and $x_k = (b - a)/2$. Since input intervals usually represent independent variables, they are assumed to be unrelated, and a new noise symbol ε_k must be used for each input interval. Conversely, the value of a quantity represented by an affine form $\hat{x} = x_0 + x_1 \varepsilon_1 + \cdots + x_n \varepsilon_n$ is guaranteed to be in the interval $[\hat{x}] = [x_0 - \xi, x_0 + \xi]$, where $\xi = \|\hat{x}\| := \sum_{i=1}^{n} |x_i|$. Note that the *range* $[\hat{x}]$ is the smallest interval that contains all possible values of $\hat{x}$ because, by definition, each noise symbol ε_i ranges independently over the interval $[-1, +1]$.

To evaluate an expression with AA, simply use the AA version of each elementary operation $z = f(x, y, \dots)$, as done in IA. Affine operations are the easiest because they can be computed exactly. Given two affine forms

$$\hat{x} = x_0 + x_1 \varepsilon_1 + \cdots + x_n \varepsilon_n$$
$$\hat{y} = y_0 + y_1 \varepsilon_1 + \cdots + y_n \varepsilon_n$$

and a scalar $\alpha \in \mathbb{R}$, we have

$$\hat{x} \pm \hat{y} = (x_0 \pm y_0) + (x_1 \pm y_1)\varepsilon_1 + \cdots + (x_n \pm y_n)\varepsilon_n$$
$$\alpha\hat{x} = (\alpha x_0) + (\alpha x_1)\varepsilon_1 + \cdots + (\alpha x_n)\varepsilon_n$$
$$\hat{x} \pm \alpha = (x_0 \pm \alpha) + x_1 \varepsilon_1 + \cdots + x_n \varepsilon_n.$$

Note that $\hat{x} - \hat{x}$ is identically zero in AA. As we saw earlier, the absence of such trivial cancellations is a major ingredient in the dependency problem in IA.

For nonaffine operations, we pick a good affine approximation for f and append an extra term $z_k\varepsilon_k$ to represent the error in this approximation:

$$\hat{z} = z_0 + z_1\varepsilon_1 + \cdots + z_n\varepsilon_n + z_k\varepsilon_k.$$

Again, ε_k is a new noise symbol and z_k is an upper bound for the approximation error. In this way, we can write AA formulas for all elementary operations and functions (for details, see Stolfi and de Figueiredo, 1997). For instance, the multiplication is given by

$$z_0 = x_0 y_0$$
$$z_i = x_0 y_i + y_0 x_i \qquad (i = 1, \ldots, n)$$
$$z_k = \|\hat{x}\|\, \|\hat{y}\|.$$

To see how AA handles the dependency problem, consider again the example given earlier: evaluate $z = x(10 - x)$ for x in the interval $[4, 6]$:

$$\hat{x} = 5 + 1\varepsilon_1$$
$$10 - \hat{x} = 5 - 1\varepsilon_1$$
$$\hat{z} = \hat{x}(10 - \hat{x}) = 25 + 0\varepsilon_1 + 1\varepsilon_2$$
$$[\hat{z}] = [25 - 1, 25 + 1] = [24, 26].$$

Note that the correlation between x and $10 - x$ was canceled in the product. Note also that the interval associated with $\hat{z}$ is much closer to the exact range $[24, 25]$ and much better than the estimate computed with IA $[16, 36]$. Expanding the expression to $10x - x^2$ makes IA give a much worse result $[4, 44]$, whereas it allows AA to compute the exact values $[24, 25]$.

Although AA is harder to implement and more expensive to execute than IA, many times the better estimates provided by AA make algorithms faster as a whole.

7.7 EXAMPLES IN COMPUTER GRAPHICS

In this section, we give examples of the use of genetic algorithms and branch-and-bound global optimization in computer graphics problems.

7.7.1 DESIGNING A STABLE TABLE

We start by describing a genetic formulation for the design of a table having a square top and four legs and that is stable (Bentley, 1999). The legs are attached to the top at positions along the diagonals of the table top, and they can have different lengths (Figure 7.5). Clearly, in the optimum configuration—the stable one—the legs are attached to the corners of the table and have the same length.

Here is a genetic model for this problem (Bentley, 1999). Individuals represent the various possible configurations. Each individual has a single chromosome with eight genes: four values to measure the distance of each leg to the center of the table and four values to measure the length of each leg. Each of these eight values is represented in 8 bits, giving a chromosome with 64 bits. Thus, the candidate space has $2^{64} \approx 2 \times 10^{19}$ elements. The objective function evaluates the stability of the table from the position and length of the leg. Starting with a random initial population, the optimum solution is reached.

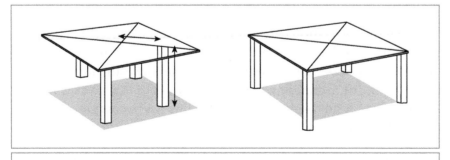

Figure 7.5: Genetic formulation for the design of a stable table (Bentley, 1999).

7.7.2 CURVE FITTING

We now consider the classic curve fitting problem: a set of points on the plane is given and we wish to find the curve that best fits them. If the search is limited to a fixed class of functions (e.g., polynomials of degree 3), then the fitting problem can be formulated as a least-squares problem and solved using continuous optimization on the parameters of the fitting model, as seen in Chapter 4. The following genetic formulation allows a fitting that is not restricted to a small class of functions (Sokoryansky and Shafran, 1997).

In this genetic model, individuals are curves whose fitness is measured by how well the curve fits the given data points. The genotype is an algebraic expression whose tree is represented linearly in LISP. For instance, the expression $t^2 - t + 1$ can be represented by the list `(+ (-(* t t) t) 1)`. This representation may be inconvenient for a human reader but is very suitable to the symbolic manipulations that are needed during the evolution. For instance, crossed reproduction takes place by simply mixing branches of the trees (see Figure 7.6).

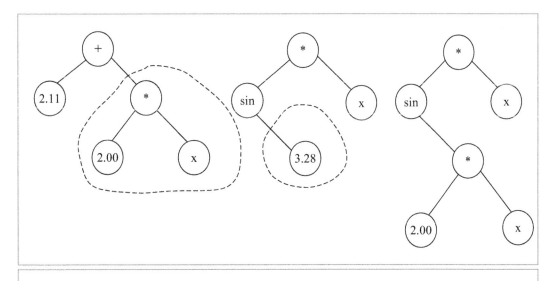

Figure 7.6: Crossed reproduction of algebraic expressions.

Figure 7.7 shows an example of curve fitting by expression evolution. The set of data points $\{x_i, y_i\}$ consists of five points sampled from the curve $y = x^2 - x + 1$. The initial population had 100 random individuals. The objective function used to measure the fitness of a curve $y = g(x)$ was

$$\sum_{i=1}^{n} \frac{|g(x_i) - y_i|^3}{y_i}.$$

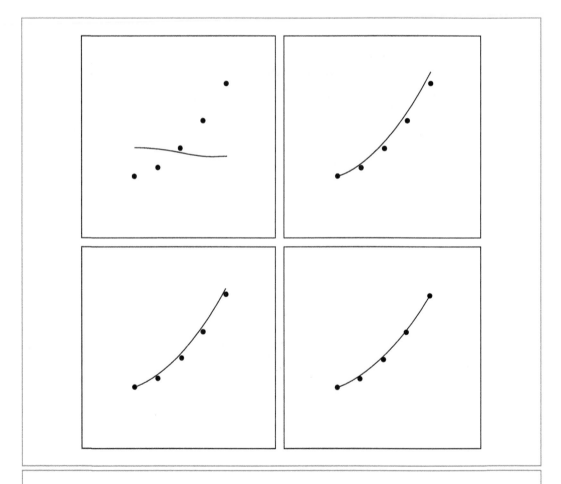

Figure 7.7: Curve fitting by evolution.

Figure 7.7 shows the best individuals in generations 1, 10, 37, and 64; their genotypes are given below:

Generation	Genotype of best individual
1	`(+ 6.15925 (cos (−0.889367 (% (−t` `(−(hsin −0.90455) −1.51327)) t))))`
10	`(* t (/ t 1.06113))`
37	`(* t (/ (/ t 1.06113) 1.06113))`
64	`(* (* (cos (sin −0.632015)) t) t).`

Note that although the final fitting is good, its expression,

$$y = x^2 \cos(\sin(-0.632015)) \approx 0.830511\,x^2,$$

does not correspond to the original expression $y = x^2 - x + 1$. It is unreasonable to expect that it is possible to evolve to the original expression because the solution space contains much more than just polynomial functions (e.g., the best individual in generation 10 is a rational function). It is even surprising that in this example the algorithm has found a polynomial of the correct degree. In any case, in practice, we do not know the original expression.

7.7.3 COLLISION DETECTION

Consider a physically based animation in which objects move subject to the action of internal and external forces, such as gravity. To achieve a realistic animation, it necessary to know how to detect when two objects collide because, at that moment, new forces appear instantaneously.

The problem of detecting the collision of two objects given parametrically by $S_1, S_2 : \mathbb{R}^2 \to \mathbb{R}^3$ can be formulated as a feasibility problem

$$S_1(u_1, v_1) - S_2(u_2, u_2) = 0,$$

for $(u_1, v_1, u_2, u_2) \in \mathbb{R}^4$. (Note that this problem has three equations, one for each coordinate in $\mathbb{R}^3$.) If the objects are given implicitly by $S_1, S_2 : \mathbb{R}^3 \rightarrow \mathbb{R}$, then the feasibility problem is

$$S_1(x, y, z) = 0, \qquad S_2(x, y, z) = 0,$$

for $(x, y, z) \in \mathbb{R}^3$.

These problems can be solved reliably using an interval branch-and-bound method. Note that it is not necessary to compute the whole feasible set in this case: it is enough to establish whether it is empty or not. When there is a collision, the search for the feasible set $\widetilde{\Omega}$ can stop as soon as a subregion is accepted as solution. This frequently happens quite early in the exploration of the domain Ω.

7.7.4 SURFACE INTERSECTION

When we compute the feasible set for detecting collisions, we actually solve a surface intersection problem. Figure 7.8 shows two parametric surfaces that intersect each other: a mechanical part and a cylinder. Figure 7.8 also shows the object obtained by removing the interior of the cylinder from the mechanical part. This removal is made by identifying the intersection curve in the parametric domains of the surfaces. Figure 7.9 shows the projection of the feasible set on each domain; these projections provide an approximation to the intersection curve that is used for trimming, as shown in Figure 7.8.

Figure 7.10 shows two other intersecting surfaces. Figure 7.11 shows the domain decompositions computed with both IA and AA. The two surfaces are Bézier surfaces—their expressions contain many correlations, which are exploited by AA when computing the intersection, yeilding a better approximation for the same tolerance.

7.7.5 IMPLICIT CURVES AND SURFACES

Another problem that is naturally described as a feasibility problem is approximating implicit objects: given $f : \Omega \subseteq \mathbb{R}^d \rightarrow \mathbb{R}$, where $d = 2$ or

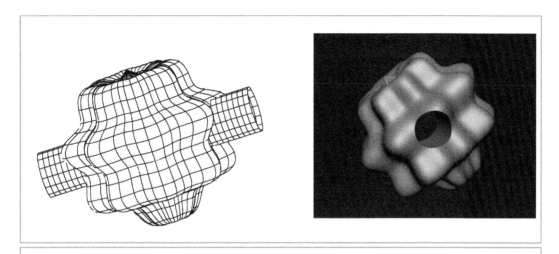

Figure 7.8: Intersection of parametric surfaces (Snyder, 1991).

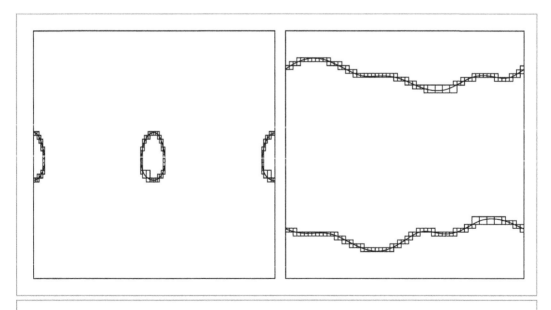

Figure 7.9: Trimming curves (Snyder, 1991).

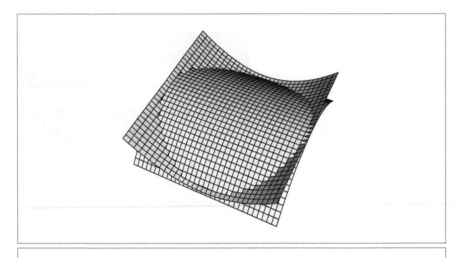

Figure 7.10: Intersection of parametric surfaces.

$d = 3$, compute an approximation to the curve or surface $S = f^{-1}(0) = \{x \in \Omega : f(x) = 0\}$.

As described earlier, this problem can be solved by interval branch-and-bound methods. Given an inclusion function F for f, the domain Ω is subdivided recursively and we discard those subregions X of Ω for which $0 \notin F(X)$. What is left is a collection of small cells that contain S. This cellular model, which has dimension d, can be used to extract an approximation of S of the right dimension, $d - 1$, by linear interpolation of the values of f at the vertices, as shown in Figure 7.12.

Figure 7.13 shows that the use of AA can provide better approximations than IA within the same tolerance. The curve shown in Figure 7.13 is the quartic given by

$$x^2 + y^2 + xy - (xy)^2/2 - 1/4 = 0.$$

Note how AA was able to exploit the correlations in the terms of this expression to compute a better approximation. IA was unable to compute an approximation with the right number of connected components.

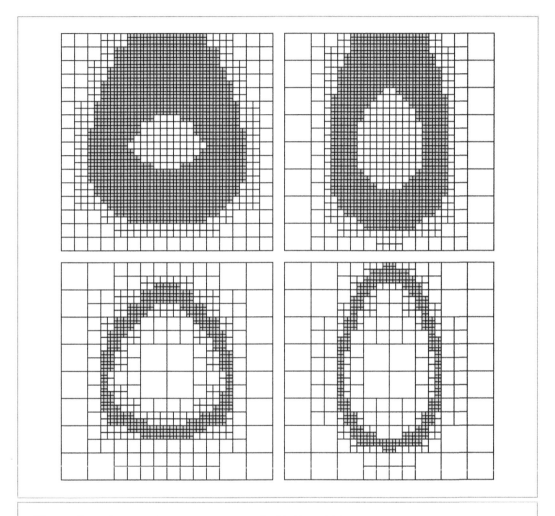

Figure 7.11: Intersection curves computed with IA and AA.

7.7.6 ADAPTIVE APPROXIMATION OF PARAMETRIC SURFACES

The simplest method for approximating a parametric surface by a polygonal mesh is to use a uniform decomposition of its domain. However,

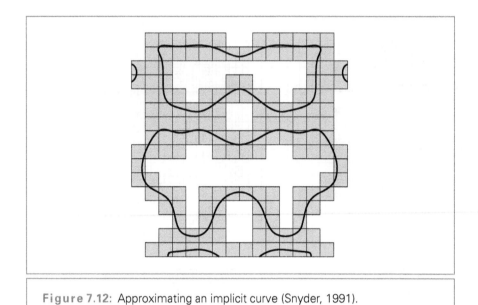

Figure 7.12: Approximating an implicit curve (Snyder, 1991).

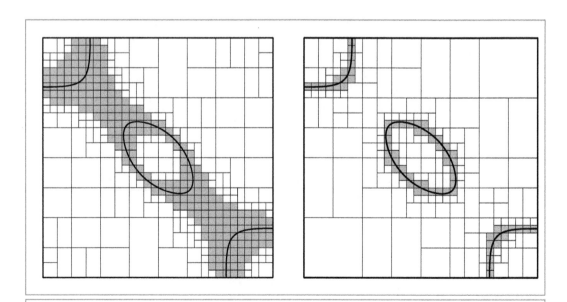

Figure 7.13: Approximating an implicit curve with IA and AA.

it is generally necessary to use a very fine mesh in the domain to capture all details of the surface. This method is inefficient and generates very large models because in general surfaces have details on several scales and it is not necessary to use a fine mesh in the whole domain.

If the goal is to compute a polygonal approximation such that the distance between any two vertices is at most at user-selected tolerance ε, then we can formulate the problem of approximating a parametric surface $S : \mathbb{R}^2 \rightarrow \mathbb{R}^3$ as a feasibility problem in which the constraints are of the form

$$|S(u, v) - S(u', v')| < \varepsilon.$$

Figure 7.14 shows a uniform decomposition and a decomposition adapted to the criterion above, along with the corresponding polygonal decompositions. Note how the points in the adapted approximation are concentrated near the details of the surface.

7.8 COMMENTS AND REFERENCES

Global optimization methods encompass different techniques to approach hard minimization problems for which it is not possible to find a global minimum efficiently by continuous or combinatorial methods.

Stochastic and genetic techniques are inspired by physics and biology to explore the solution space with the hope to find the global minimum.

Simulated annealing is one of the best known stochastic techniques for global optimization. A seminal paper on this subject that presents the Metropolis algorithm is that by Metropolis *et al.* (1953). Another relevant paper is that of Kirkpatrick *et al.* (1983).

We remark that there is a connection between stochastic optimization techniques and probability theory applied to optimization. We discuss this issue in the next chapter.

Figure 7.14: Approximating parametric surfaces (Snyder, 1991).

Genetic algorithms have been extensively used in graphics and vision. Besides the examples discussed in Section 7.4.2, it is worth mentioning the work of Karl Sims in animation (Sims, 1994) and image synthesis (Sims, 1991b).

BIBLIOGRAPHY

Baker Kearfott, R. *Rigorous Global Search: Continuous Problems.* Dordrecht, Netherlands: Kluwer, 1996.

Beasley, D., D. R. Bull, and R. R. Martin. An overview of genetic algorithms: Part 1, fundamentals. *University Computing,* 15(2):58–69, 1993.

Bentley, P. Aspects of evolutionary design by computers. *Advances in Soft Computing - Engineering Design and Manufacturing.* London, Springer-Verlag, 99–1180, 1999.

Berner, S. New results on verified global optimization. *Computing,* 57(4):323–343, 1996.

Comba, J. L. D. and J. Stolfi. Affine arithmetic and its applications to computer graphics. In *Anais do VI SIBGRAPI,* 9–18, 1993.

Csendes, T., and D. Ratz. Subdivision direction selection in interval methods for global optimization. *SIAM Journal of Numerical Analysis,* 34(3):922–938, 1997.

de Figueiredo, L. H. and Stolfi, J. Affine arithmetic: concepts and applications. Numerical Algorithms 37:1–4, 147–158, 2004.

Hansen, E. *Global Optimization using Interval Analysis.* Monographs and Textbooks in Pure and Applied Mathematics, no. 165. New York: Marcel Dekker, 1988.

Holland, J. H. *Adaptation in Natural and Artificial Systems.* An Introductory Analysis with Applications to Biology, Control, and Artificial Intelligence, Ann Arbor, MI: University of Michigan Press, 1975.

Kirkpatrick, S., C. D. Gelatt, and M. P. Vecchi. Optimization by simulated annealing. *Science,* 220(4598):671–680, 1983.

Koza, J. R. *Genetic Programming: On the Programming of Computers by Means of Natural Selection.* Cambridge, MA: MIT Press, 1992.

Kreinovich, V. *Interval Software.*

Metropolis, N., A. W. Rosenbluth, M. N. Rosenbluth, A. H. Teller, and E. Teller. Equations of state calculations by fast computing machines. *Journal of Chemical Physics,* 21(6):1087–1092, 1953.

Michalewicz, Z. *Genetic Algorithms + Data Structures = Evolution Programs*. Second edn. Berlin: Springer-Verlag, 1994.

Moore, R. E. *Interval Analysis*. Amsterdam, The Netherlands: Prentice-Hall, 1966.

Moore, R. E. *Methods and Applications of Interval Analysis*. Philadelphia, PA: SIAM, 1979.

Moore, R. E. Global optimization to prescribed accuracy. *Computers and Mathematics with Applications*, 21(6–7):25–39, 1991.

Ratschek, H., and J. Rokne. *Computer Methods for the Range of Functions*. Chichester, UK: Ellis Horwood Ltd, 1984.

Ratschek, H., and J. Rokne. *New Computer Methods for Global Optimization*. Chichester, UK: Ellis Horwood Ltd, 1988.

Rump, S. M. Algorithms for verified inclusions: theory and practice. In R. E. Moore (Ed), *Reliability in Computing: The Role of Interval Methods in Scientific Computing. Perspectives in Computing*, vol. 19. Boston, MA: Academic Press, 109–126, 1988.

Sims, K. Artificial evolution for computer graphics. *Computer Graphics (SIGGRAPH '91 Proceedings)*, 25(July): 319–328, 1991a.

Sims, K. Artificial evolution for computer graphics. *Computer Graphics (Proceedings of SIGGRAPH 91)*, 319–328, 1991b (July).

Sims, K. Evolving virtual creatures. In *Proceedings of SIGGRAPH 94. Computer Graphics Proceedings*, Annual Conference Series, NY: ACM Press, 15–22, 1994 (July).

Snyder, J. M. *Generative Modeling: An Approach to High Level Shape Design for Computer Graphics and CAD*. Ph.D. Thesis, California Institute of Technology, 1991.

Sokoryansky, M., and A. Shafran. *Evolutionary Development of Computer Graphics*, 1997 (Fall).

Stolfi, J., and L. H. de Figueiredo. *Self-Validated Numerical Methods and Applications*. 21º Colóquio Brasileiro de Matemática, IMPA, 1997.

8 PROBABILITY AND OPTIMIZATION

In this chapter, we present the basic principles of the information theory. We discuss the concept of entropy in thermodynamics and the optimization principles that are based on entropy.

8.1 BACKGROUND

We begin by exploring the relationships between probability and the degree of organization of a physical system.

8.1.1 THE SECOND LAW OF THERMODYNAMICS

The science of thermodynamics began in the first part of the nineteenth century with the study of gases, with a motivation to improve the efficiency of the steam engines. Nonetheless, the principles of thermodynamics are valid for any physical systems.

Thermodynamics is an experimental science based on a small number of principles, which are generalizations made from experiments. These laws are concerned only with the macroscopic properties of matter (e.g., large scale). Thermodynamics play a fundamental role in physics and their principles continue valid, even after the advance of physics in the twentieth century.

In the first part of the nineteenth century, researchers observed that various processes occur in a spontaneous way. For example, if two bodies of different temperatures are in contact with each other, these bodies will exchange their temperatures such that they both will have the same temperature. The converse behavior (i.e., two bodies with the same temperature in contact and one body gets hotter and the other cooler) never occurs. The second law of thermodynamics determines the direction in which a process occurs. In short, it can be stated as follows: *the entropy in the universe always increases.*

The second law of thermodynamics states that in any process that occurs (in time) within an isolated system (i.e., which does not exchange mass or energy with the exterior), the entropy does not decrease. This is the only physical law that prescribes a preferential direction for the time domain. For that reason, it is of fundamental importance in physics, as we can verify from the words of the physicists Arthur Eddington and Albert Einstein.

> The law that entropy always increases—the second law of thermodynamics—holds, I think, the supreme position among the laws of Nature. If someone points out to you that your pet theory of the universe is in disagreement with Maxwell's equations—then so much worse for Maxwell equations. If it is found to be contradicted by observation—well these experimentalists do bungle things sometimes. But if your theory is found to be against the second law of Thermodynamics, I can give you no hope; there is nothing for it but to collapse in deepest humiliation.
>
> —Arthur Eddington (1948)

> [A law] is more impressive the greater the simplicity of its premises, the more different are the kinds of things it relates, and the more extended its range of applicability. Therefore, the deep impression which classical thermodynamics made on me. It is the only physical theory of universal content, which I am

convinced, that within the framework of applicability of its basic concepts will never be overthrown.

—Albert Einstein Klein (1967)

In thermodynamics, the variation of entropy is defined as the heat variation over the temperature

$$dS = \frac{dQ}{T}.$$

From statistical thermodynamics, we have the Boltzmann formula that defines the entropy of a physical system

$$S = k \log W,$$

where S is the entropy, $k = 1.381 \times 10^{-16}$ is the Boltzmann constant, and W is the number of accessible microstates (they are the possible values that determine the state of each particle in the system). Quantum mechanics says that W is a finite integer number.

Boltzmann himself noticed that the entropy measures the lack of organization in a physical system. The identification of information entropy was clarified later with the explanation of Szilard to the Maxwell devil paradox (described in the next paragraph) and also with the information theory of Shannon.

Maxwell proposed the following. Let a chamber divided in two parts A and B be separated by a partition (which does not transfer heat), a small moving door. In each part, we have samples of a gas at the same temperature. The gas temperature is determined by a statistical "mean" of the velocities of the gas particles. However, in a gas sample, there are slow and fast particles (with lower and higher velocities than the statistical mean, respectively). A devil will control the door between A and B, letting only slow particles from A to B and letting only fast particles from B to A. With this, the temperature of A will increase and the temperature of B will decrease, which is not according to our intuition and also contradicts the second law of thermodynamics.

Szilard[1] in 1929 proposed the following explanation for the Maxwell devil paradox. The devil must be very well informed about the velocity of all gas particles. Szilard identified information with entropy and showed that the available information increases the entropy of the system chamber and devil. The idea of his argument is that from quantum mechanics, the energy is quantized. By associating a packet of energy with a bit of information about a particle, he calculated the total entropy of the system chamber and devil and showed that the entropy increases. This argument demonstrates that any system that manipulates information (e.g., a brain or a computer) can be considered a thermomachine, and the entropy always increases in the system. According to Resnikoff (1989), a computer (in the '80s) would dissipate energy in the order of 10^{10} times this fundamental physical limit.

8.1.2 INFORMATION AND MEASURES

Before presenting the information theory in the original form described by Shannon (1948), we comment on the gain of information as a measure.

We wish to measure the value of a variable x. Let us assume that the measure results in an interval $[a, b]$. We have a previous measure of x corresponding to the interval $[0, 1]$. From the new measure, we obtain that x belongs to an interval Δx (contained in $[0, 1]$). What is the gain of information with this new measure?

We can answer this question, considering the space (in bits, or symbols) required to represent the gain of this new measure. To make it simpler, let us suppose that $0 < x < 1$ is a number in decimal format, e.g., $x = 0,23765\ldots$, and that for each new measure, we obtain new digits of the decimal representation of x. For example, in the first measure, we obtain 23, and in the second measure, we obtain 765. The space necessary to represent the information gain of the first and second measures is two and three digits, respectively. Thus, we conclude that for an

---

1   A German physicist, who was involved in the first experiment on a nuclear reaction in 1942 in the University of Chicago.

information gain of two digits, it is necessary to divide the measure interval by 100. For an information gain of three digits, it is necessary to divide it by 1000, and so on. With a little more attention to the arguments above and with the inclusion of some "desirable" hypotheses for the information gain of a measure, it is possible to demonstrate that this information gain is

$$I = -\log\left(\frac{\Delta x_2}{\Delta x_1}\right),$$

where Δx_2 is the interval of the new measure and Δx_1 is the interval of the previous measure. In a binary system, the information is expressed by the number of bits and the base of the logarithm should be 2. For the information in decimal digits, the logarithm is in base 10.

Suppose we observe a system with a finite number of states $S_1, S_2, \ldots, S_n$. Each state S_i has an occurrence probability p_i, with

$$p_1 + p_2 + \cdots + p_n = 1.$$

The gain of information with the observation of the occurence in state S_i is $-\log\left(\frac{\Delta x_2}{\Delta x_1}\right) = -\log\left(\frac{p_i}{1}\right) = -\log p_i$. Therefore, the expectation on the information gain is

$$S = -\sum_{i=1}^{n} p_i \log p_i.$$

8.2 INFORMATION THEORY

Information theory relates the degree of organization of a system to information.

8.2.1 BASIC PRINCIPLES

Information theory began with Shannon in 1948. Shannon's starting point for the formula of entropy was axiomatic. He imposed that the entropy (or information) had some properties and demonstrated that there is only one possible formula for the gain of information under these assumptions. Below we show a small excerpt from his famous paper (Shannon, 1948), where he describes the desirable properties for the entropy.

> Suppose we have a set of possible events whose probabilities of occurrence are $p_1, p_2, \ldots, p_n$ $(p_1 + p_2 + \cdots + p_n = 1)$. These probabilities are known but that is all we know concerning which event will occur. Can we find a measure of how much choice is involved in the selection of the event or of how uncertain we are of the outcome?

If there is such a measure, say $H(p_1, p_2, \ldots, p_n)$, it is reasonable to impose of it the following properties:

1. H should be continuous in the p_i.
2. If all the p_i are equal, $p_i = \frac{1}{n}$, then H should be a monotonic increasing function of n. With equally likely events, there is more choice or uncertainty when there are more possible events.
3. If a choice is broken down into two successive choices, the original H should be the weighted sum of the individual values of H, which is illustrated in Figure 8.1. On the left, we have three possibilities $p_1 = \frac{1}{2}, p_2 = \frac{1}{3}$, and $p_3 = \frac{1}{6}$. On the right, we first choose between two possibilities, each with a probability $\frac{1}{2}$, and if the second occurs, make another choice with probabilities $\frac{2}{3}$ and $\frac{1}{3}$. The final results have the same probabilities as before. We require, in this special case, that

$$H\left(\frac{1}{2}, \frac{1}{3}, \frac{1}{6}\right) = H\left(\frac{1}{2}, \frac{1}{2}\right) + \frac{1}{2}H\left(\frac{2}{3}, \frac{1}{3}\right).$$

The coefficient $\frac{1}{2}$ is because this second choice only occurs half the time.

Property 3 indicates that if, for example, we have NL symbols with the same probability and we group these NL symbols into N groups of L symbols, and if the information for an output symbol is given in two steps, first indicating the group and second the specific element of this

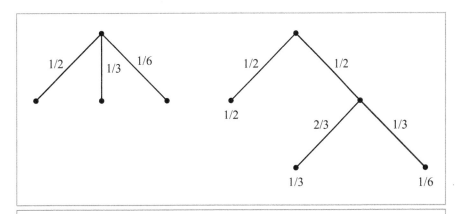

Figure 8.1: Decomposition from the choice of three possibilities.

group, then the total information is the sum of information about the group and the symbol (inside the group), i.e., $I(NL) = I(N) + I(L)$.

Shannon demonstrated that the only H that satisfies these three properties is of the form

$$H = -k \sum_i p_i \log p_i,$$

where k is a positive constant, and without loss of generality, we can consider $k = 1$.

From now on, we denote this quantity by *entropy*, whose formula is

$$S = -\sum_{i=1}^{n} p_i \log p_i.$$

Figure 8.2 shows the graph of the entropy in bits for two random variables. That is, $S = -p \log_2 p - (1 - p) \log_2(1 - p)$.

8.2.2 PROPERTIES OF ENTROPY

Before we present the main properties of entropy, we need to establish the following notation:

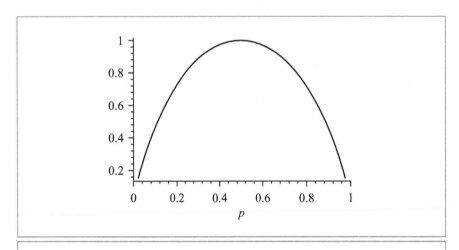

Figure 8.2: Graph of $S = -p\log_2 p - (1 - p)\log_2(1 - p)$.

Let $\mathbf{p} = (p_1, p_2, \ldots, p_n)$ be a probability distribution with $p_i \geqslant 0$ and $\sum_{i=1}^{n} p_i = 1$.

We denote the entropy as

$$S_n(\mathbf{p}) = -\sum_{i=1}^{n} p_i \log p_i.$$

Since $p_i \log p_i$ is not defined for $p_i = 0$, we take $0 \log 0 = \lim_{p \to 0} p \log p = 0$.

1. For the n degenerate distributions $\Delta_1 = (1, 0, \ldots, 0)$, $\Delta_2 = (0, 1, \ldots, 0), \ldots, \Delta_n = (0, 0, \ldots, 1)$, we have $S_n(\Delta_i) = 0$, which is desirable because in this case, there is no uncertainty in the output.

2. For any probability distribution that is not degenerate, we have $S_n(\mathbf{p}) > 0$, which is desirable because there is uncertainty in the output.

3. $S_n(\mathbf{p})$ does not change if we permute in $p_1, p_2, \ldots, p_n$. This property is desirable because the order of $p_i's$ in the probability distribution cannot change the entropy.

4. The entropy does not change with the inclusion of an impossible event, i.e.:

$$S_{n+1}(p_1, p_2, \ldots, p_n, 0) = S_n(p_1, p_2, \ldots, p_n).$$

5. $S_n(\mathbf{p})$ is a concave function with maximum at $p_1 = p_2 = \cdots = p_n = \frac{1}{n}$. In this case, the value of the entropy is

$$S_n(\mathbf{p}) = -\sum_{i=1}^{n} p_i \log p_i = -n \left(\frac{1}{n} \log \frac{1}{n} \right) = \log n.$$

8.3 MEASURING MUTUAL INFORMATION

In order to put probability into the context of optimization, we need to present the concepts of conditional entropy, mutual information, and divergence.

8.3.1 CONDITIONAL ENTROPY

Before we formulate the concept of conditional entropy, we introduce the notation for random variables.

We can consider the output of an information source as the realization of a random variable, denoted by X.

The random variable X is denoted by the alphabet of possible values X can assume: $\{x_1, x_2, \ldots, x_n\}$ and by its probability distribution $\mathbf{p} = (p_1, p_2, \ldots, p_n)$, where $p_i = P\{X = x_i\}$.

The expectation of X is $E[X] = \sum_{i=1}^{n} x_i p_i$.

We denote by $S(X)$ the entropy governed by the distribution $\mathbf{p}$.

With this notation, we consider a new random variable $Z(X) = -\log P\{X\}$, thus $E[Z] = E\left[-\log P\{X\}\right] = -\sum_{i=1}^{n} p_i \cdot \log p_i = S(X)$.

From this formula, we have that the entropy is a mean and the value $-\log p_x$ is the quantity of information supplied by the output of x.

The entropy corresponding to the output sequence of x, y is

$$S(X, Y) = -\sum_{x,y} P\{X = x, Y = y\} \log P\{X = x, Y = y\}.$$

If the source is independent of time and memoryless, we have

$$S(X_1, X_2, \ldots, X_n) = nS(X).$$

It can be proved that the entropy is maximal if the random variables are independent. In general, we are interested in the entropy per output unity, defined as

$$S = \frac{1}{k} \lim_{k \to \infty} S(X_1, X_2, \ldots, X_k).$$

The definition of conditional entropy is

$$S(X|Y) = -\sum_{x,y} P\{X = x, Y = y\} \log P\{X = x|Y = y\}.$$

We have $S(X|Y) = -\sum_{y} P\{Y = y\} S\{X|Y = y\}$, i.e., the conditional entropy is an average of entropies.

An important relation of conditional entropy is that

$$S(X, Y) = S(X) + S(Y|X).$$

This equation can be interpreted in the following way: the uncertainty of two random variables is equal to the uncertainty of the first variable and the mean uncertainty of the second variable with the first variable known.

The last property of conditional entropy is

$$S(X_1|X_2,\ldots,X_k,X_{k+1}) \leqslant S(X_1|X_2,\ldots,X_k).$$

That is, the uncertainty decreases as more information is known.

8.3.2 MUTUAL INFORMATION

The mutual information between two random variables X and Y is defined as

$$I(X;Y) = S(X) - S(X|Y).$$

We have

$$I(X;Y) = S(X) - S(X|Y) = \sum_{x,y} P(x,y) \log \frac{P(x|y)}{P(x)}$$
$$= \sum_{x,y} P(x,y) \log \frac{P(x,y)}{P(x)P(y)} = S(X) + S(Y) - S(X,Y) = I(Y;X).$$

The above equation tells us that the mutual information is symmetric and also that

$$0 \leqslant I(X;Y) \leqslant \min \{S(X), S(Y)\}.$$

We have $I(X;Y) = 0$ if and only if X and Y are independent and $I(X;Y) = \min\{S(X), S(Y)\}$ if and only if X determines Y or Y determines X.

Finally, we note that

$$I(X;Y) = \sum_{x,y} P(x,y) \log \frac{P(x,y)}{P(x)P(y)} = E\left[\log \frac{P(x,y)}{P(x)P(y)}\right].$$

We can interpret $\log \frac{P(x,y)}{P(x)P(y)}$ as the mutual information between the symbols x and y.

8.3.3 DIVERGENCE

Often it is necessary to measure the distance between two probability distributions $\mathbf{p}$ and $\mathbf{q}$. The most common measure was developed by Kullback and Leibler in 1951 and is known as distance (or divergence or measure) of Kullback–Leibler, relative entropy, discriminant, or simply divergence.

The divergence is defined as

$$D(\mathbf{p}; \mathbf{q}) = \sum_{i=1}^{n} p_i \cdot \log \frac{p_i}{q_i},$$

where we assume that if $q_i = 0$, then $p_i = 0$ and that $0 \log \frac{0}{0} = 0$.

The divergence is not symmetric, i.e., $D(\mathbf{p}; \mathbf{q}) \neq D(\mathbf{q}; \mathbf{p})$ in general. Eventually, we can use the symmetric divergence, defined by

$$J(\mathbf{p}; \mathbf{q}) = D(\mathbf{p}; \mathbf{q}) + D(\mathbf{q}; \mathbf{p}).$$

As the entropy, the divergence has several interesting properties:

1. $D(\mathbf{p}; \mathbf{q}) = 0$ only if $\mathbf{p} = \mathbf{q}$. If $\mathbf{p} \neq \mathbf{q}$, then $D(\mathbf{p}; \mathbf{q}) > 0$.
2. $D(\mathbf{p}; \mathbf{q})$ is continuous and convex in $\mathbf{p}$ and $\mathbf{q}$.
3. $D(\mathbf{p}; \mathbf{q})$ is permutationally symmetric, i.e., $D(\mathbf{p}; \mathbf{q})$ does not alter if the pairs $(p_1, q_1), \ldots, (p_n, q_n)$ are permuted.
4. If $\mathbf{U} = \left(\frac{1}{n}, \frac{1}{n}, \ldots, \frac{1}{n} \right)$ is the uniform distribution, then

$$D(\mathbf{p}; \mathbf{U}) = \sum_{i=1}^{n} p_i \cdot \log \frac{p_i}{1/n} = \log n + \sum_{i=1}^{n} p_i \cdot \log p_i = \log n - S_n(\mathbf{p}).$$

5. $I(X; Y) = \sum_{x,y} P(x, y) \log \frac{P(x,y)}{P(x)P(y)} = D(\mathbf{p}; \mathbf{q})$, where $\mathbf{p} = P(x, y)$ and
   $\mathbf{q} = P(x)P(y)$.

8.3.4 CONTINUOUS RANDOM VARIABLES

In this section, we give definitions of entropy, conditional entropy, divergence, and mutual information for continuous random variables. They are similar to the discrete case, with the basic difference of changing the symbol $\sum$ by $\int$.

Let X be a continuous random variable, with probability distribution $f_X(x)$. The entropy of X is called differential entropy and is defined by

$$S(X) = -\int f_X(x) \log f_X(x) dx = -E\big[\log f_X(x)\big].$$

The differential entropy has some of the properties of the ordinary entropy (for discrete variables); however, it can be negative and it is not an absolute measure of the randomness of X.

The mutual information of continuous variables has the same definition as in the discrete case:

$$I(X; Y) = S(X) - S(X|Y),$$

where the conditional entropy is defined by

$$S(X|Y) = -\int\int f_{X,Y}(x, y) \log f_X(x|y) dx dy.$$

The relation below is valid:

$$I(X; Y) = \int\int f_X(x, y) \log \frac{f_X(x|y)}{f_X(x)} dx dy$$
$$= \int\int f_{X,Y}(x, y) \log \frac{f_{X,Y}(x, y)}{f_X(x)f_Y(y)} dx dy.$$

The divergence is defined by:

$$D\big(f_X; g_X\big) = \int f_X(x) \log \frac{f_X(x)}{g_X(x)} dx.$$

The divergence for continuous variable has some unique properties:

- $D\big(f_X; g_X\big) = 0$ only if $f_X = g_X$. If $f_X \neq g_X$, then $D\big(f_X; g_X\big) > 0$.
- The divergence is invariant in relation to the following variations in the components of the vector **x**: permutations in the order of components, scaling in amplitude, and monotonous nonlinear transformations.

Also, we have $I(X; Y) = D\big(f_{X,Y}; f_X f_Y\big)$.

8.4 OPTIMIZATION AND ENTROPY

In this section, we discuss optimization principles that are based on entropy.

8.4.1 PRINCIPLE OF MAXIMUM ENTROPY—MaxEnt

In nature, due to the second law of thermodynamics, the entropy of a system always increases (or better, it never decreases). However, in general, the entropy of a physical system tends to the distribution of maximum entropy, respecting the physical restrictions that the system imposes.

The principle of maximum entropy is based on this idea, i.e., to find the distribution of maximum entropy, respecting the imposed restrictions.

Let X be a random variable that can assume the values $x_1, x_2, \ldots, x_n$, but with the corresponding probabilities $p_1, p_2, \ldots, p_n$ unknown. However, we know some restrictions about the probability distributions, such as the mean, variance, and some moments of X. How to choose the probabilities $p_1, p_2, \ldots, p_n$?

This problem, in general, has infinite solutions. We have to look for a function to be optimized such that we find a unique solution. The principle of maximum entropy, initially proposed by Jaynes in 1957, offers a solution to this problem.

The principle of maximum entropy, which we will call MaxEnt, states that from the probability distributions that satisfy the restrictions, we have to choose the one with maximum entropy.

The MaxEnt is a very general principle and has applications in various fields. The first applications were in statistical mechanics and thermodynamics. There are applications of MaxEnt in many areas, such as urban planning, economy, queuing theory, spectral analysis models for population growth, and language. In graphics and vision, it can be applied to pattern recognition, image processing, and many other problems. An important application in statistical themodynamics is to find the probability distribution that describe the (micro)states of a system, based on measures (average values of functions that are restrictions to the problem) done in macroscopic scale.

An intuitive explanation for the MaxEnt is that if we choose a distribution with less entropy than the maximum, this reduction in entropy comes from the use of information that is not directly provided but somehow used in the problem inadvertently. For example, if we do not have any restrictions, the distribution chosen by the MaxEnt is the uniform distribution. If for some reason, we do not choose an uniform distribution, i.e., if for some i and j, we have $p_i > p_j$, then some information not provided was used to cause this asymmetry in the probability distribution.

MaxEnt can be formulated mathematically as follows. Let X be a random variable with alphabet $\{x_1, x_2, \ldots, x_n\}$ and with probability distribution $\mathbf{p} = (p_1, p_2, \ldots, p_n)$. We want to maximize the measure of Shannon subject to several linear restrictions.

$$
\begin{cases}
\arg \max H(X) = -\displaystyle\sum_{i=1}^{n} p_i \log p_i \\[2ex]
\text{subject to } p_i \geqslant 0, \forall i \text{ and } \displaystyle\sum_{i=1}^{n} p_i = 1 \text{ and } \displaystyle\sum_{i=1}^{n} p_i g_{ri} = a_r, r = 1, 2, \ldots, m.
\end{cases}
$$

Because the domain is convex and $H(X)$ is concave, we have a problem of convex optimization, which guarantees the existence of a unique solution and efficient computational methods to solve the problem.

The above problem has a specific structure that guarantees that in the solution, we have $p_i \geqslant 0, \forall i$ (because p_i are exponentials, as we will see next). In this way, we do not have to worry about inequality restrictions.

Applying the Karush–Kuhn–Tucker conditions (or Lagrange multipliers), we have the solution

$$
p_i = \exp(-\lambda_0 - \lambda_1 g_{1i} - \lambda_2 g_{2i} - \lambda_m g_{mi}), \qquad i = 1, 2, \ldots, n,
$$

where $\lambda_0, \lambda_1, \lambda_2, \ldots, \lambda_m$ are the Lagrange multipliers, which can be calculated with the equations

$$
\begin{cases}
\displaystyle\sum_{i=1}^{n} \exp\left(-\lambda_0 - \sum_{j=1}^{m} \lambda_j g_{ji}\right) = 1 \\[3ex]
\displaystyle\sum_{i=1}^{n} g_{ri} \exp\left(-\lambda_0 - \sum_{j=1}^{m} \lambda_j g_{ji}\right) = a_r, \quad r = 1, 2, \ldots, m.
\end{cases}
$$

The formulation of MaxEnt for continuous random variables is analogous to the discrete case. Consider a continuous random variable x and its continuous probability $f(x)$. If we know only the expected values $E[g_1(x)] = a_1, E[g_2(x)] = a_2, \ldots, E[g_m(x)] = a_m$ of x, we have, in general, infinite distributions satisfying these moments. According to MaxEnt, we have to choose the distribution with maximum entropy, which is

$$f(x) = \exp(-\lambda_0 - \lambda_1 g_1(x) - \lambda_2 g_2(x) - \lambda_m g_m(x)),$$

where $\lambda_0, \lambda_1, \lambda_2, \ldots, \lambda_m$ are the Lagrange multipliers, which can be determined by the following equations:

$$\begin{cases} \displaystyle\int f(x)dx = 1 \\ \displaystyle\int f(x)g_r(x)dx = a_r, \quad r = 1, 2, \ldots, m. \end{cases}$$

For more details, see Kapur and Kesavan (1992).

8.4.2 DISTRIBUTIONS OBTAINED WITH MaxEnt

Next we give the domain of a random variable and the known moments. With the application of MaxEnt and using the formulas from the previous section, either in the continuous or in the discrete case, we obtain a probability distribution with maximum entropy (PDME).

A large part of the distributions in statistics can be characterized from certain prescribed moments and the application of MaxEnt (or MinxEnt, discussed in Section 8.4.3).

1. If the domain of x is $[a, b]$ and there is no restriction (except for the natural restrictions of probability distributions), the PDME is the uniform distribution. Thus, the uniform distribution is characterized by the absence of restrictions in the entropy.
2. If the domain of x is $[a, b]$ and we have an arithmetic mean $m = E[x]$, the PDME is the truncated exponential distribution given by

$$f(x) = ce^{-kx},$$

where c and k can be calculated using the formula

$$c \int_a^b e^{-kx} dx = 1 \quad \text{and} \quad c \int_a^b x e^{-kx} dx = m.$$

3. If the domain of x is $[0, \infty)$ and we have the mean $m = E[x]$, the PDME is the distribution given by

$$f(x) = \frac{1}{m} e^{-x/m},$$

and the maximum entropy is

$$S_{\text{MAX}} = 1 + \ln m.$$

4. If the domain of x is $[0, \infty)$ and we have the arithmetic mean $m = E[x]$ and the geometric mean $E[\ln x]$, then the PDME is the gamma distribution given by

$$f(x) = \frac{a^\gamma}{\Gamma(\gamma)} e^{-ax} x^{\gamma-1}.$$

5. If the domain of x is $(-\infty, \infty)$ and we have the mean $m = E[x]$ and variance $E[(x - m)^2] = \sigma^2$, then the PDME is the normal distribution given by

$$f(x) = \frac{1}{\sigma\sqrt{2\pi}} \exp\left(-\frac{1}{2}\left(\frac{x - m}{\sigma}\right)^2\right),$$

whose entropy is

$$S_{\text{MAX}} = \ln \sigma\sqrt{2\pi e}.$$

6. If the domain of x is $(-\infty, \infty)$ and we have the moment $E[x]$, the PDME is the Laplace distribution given by

$$f(x) = \frac{1}{\sigma} \exp\left(-\frac{|x|}{\sigma}\right).$$

7. An important application in statistical thermodynamics is to find the probability distribution that describes the (micro)states of a system based on measures in microscopic scale.

Let $p_1, p_2, \ldots, p_n$ be the probabilities of the particles in the system with energies $\epsilon_1, \epsilon_2, \ldots, \epsilon_n$, respectively.

The only information we have about the system is its average energy $\widehat{\epsilon}$,

$$p_1\epsilon_1 + p_2\epsilon_2 + \cdots p_n\epsilon_n = \widehat{\epsilon},$$

and the natural restrictions of probabilities,

$$p_i \geqslant 0, \forall i \text{ and } \sum_{i=1}^{n} p_i = 1.$$

According to MaxEnt, we obtain the Maxwell–Boltzmann distribution of statistical mechanics:

$$p_i = e^{-\mu\epsilon_i} \bigg/ \left(\sum_{i=1}^{n} e^{-\mu\epsilon_i} \right), \quad i = 1, 2, \ldots, n,$$

where $\mu = \frac{1}{kT}$ (T is the temperature and k is the Boltzmann constant). Other distributions in the statistical mechanics, such as the Bose–Einstein or the Fermi–Dirac, are also obtained from the MaxEnt, changing only the (macroscopic) restrictions that we have about the system.

8.4.3 PRINCIPLES OF OPTIMIZATION BY ENTROPY

The MaxEnt is the most important optimization principle derived from entropy. However, there are many other optimization principles based on entropy, depending on the problem at hand. Kapur and Kesavan (1992) have given a general description of these principles. We briefly present the second most important principle of optimization based on entropy: the principle of mininum crossed entropy, which we call MinxEnt.

The MinxEnt is used when we do not know the probability distribution **p** but we have (as in the case of MaxEnt) restrictions relative to **p** and also an a priori probability **q** and we want to choose **p** satisfying the restrictions and be as close as possible to **q**.

The MinxEnt can be concisely enunciated as follows: from all the distributions satisfying the given restrictions, choose the one that is closest to the a priori distribution (prior). The usual distance measure for probability distributions is the divergence. With this measure, the MinxEnt says that we have to minimize the divergence $D(\mathbf{p}; \mathbf{q})$ subject to the restrictions of the problem.

Beacuse the divergence $D(\mathbf{p}; \mathbf{q})$ is convex and the domain (in general) is convex, we have (as in the case of MaxEnt) a problem of convex optimization, which guarantees the existence of a unique solution and computational efficiency.

An interesting fact is that if the prior distribution **q** is not given, the most natural choice is to pick the probability distribution **p** as close as possible to the uniform distribution $\mathbf{U} = \left(\frac{1}{n}, \frac{1}{n}, \ldots, \frac{1}{n}\right)$. As we have already seen in the section about divergence, we have

$$D(\mathbf{p}; \mathbf{U}) = \sum_{i=1}^{n} p_i \cdot \log \frac{p_i}{1/n} = \log n + \sum_{i=1}^{n} p_i \cdot \log p_i = \log n - H(\mathbf{p}).$$

Therefore, we have that to minimize the divergence in relation to the uniform distribution is equivalent to maximize the entropy. Thus, MaxEnt is a particular case of MinxEnt in which we minimize the distance in relation to the uniform distribution.

We can unite MaxEnt and MinxEnt as a general principle. "From all probability distributions satisfying the given restrictions, we have to choose the one closest to a given prior distribution. If this prior distribution is not given, choose the one closest to the uniform distribution."

8.5 APPLICATIONS

We now give examples of applications of optimization to symbol coding and decision trees.

8.5.1 SYMBOL CODING

An interesting application of the information theory is the problem of optimal signal coding. Let an output alphabet be $\alpha = \{A, B, \ldots, Z\}$ and we know the probability of the symbols $p_A, p_B, \ldots, p_Z$. What is the minimum number of bits that are to be used to encode the set of symbols of the output of some information source?

In order to achieve the minimum number of bits, it is necessary to use a variable-length code for each symbol, with less bits assigned to the most frequent symbols. Furthermore, we would like to avoid separators between the symbols (which would be an extra symbol).

A concrete example is given next. We illustrate the symbol coding problem with a source that only emits the symbols A, B, C, and D with the respective probabilities $\frac{1}{2}, \frac{1}{4}, \frac{1}{8}$, and $\frac{1}{8}$. In Table 8.1, we have two possible encodings.

A sufficient condition for a codification to be uniquely decoded is that the code of a symbol should not be a prefix of the code for any

Table 8.1: Codes α and β of the symbols A, B, C, and D.

Symbol	Probability	Encoding α	Encoding β
A	1/2	1	00
B	1/4	01	01
C	1/8	001	10
D	1/8	000	11

other symbol. This is the case of the example above. A codification that satisfies this condition can be structured in the form of a tree. In Figure 8.3, we have code trees α and β.

Let $l(x)$ be the size of the code for symbol x and $p(x)$ its probablility of ocurrence. The average length of a codification can be calculated using the formula:

$$L = \sum_x l(x)p(x).$$

Shannon (1948) demonstrated in his fundamental theorem for noiseless channels that there is no codification such that its average length L is less than the entropy $S(X)$. Furthermore, given $\varepsilon > 0$, there is a codification such that $L < S(X) + \varepsilon$.

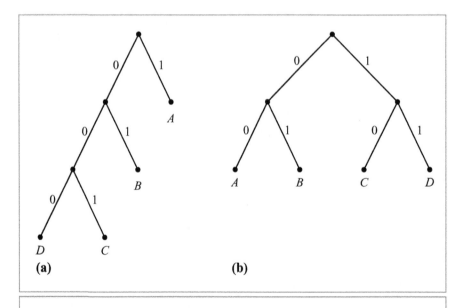

Figure 8.3: Codification trees (a) α (b) β.

The average length of the codes α and β of the example above and the associated entropy are

$$L_\alpha = \frac{1}{2} \cdot 1 + \frac{1}{4} \cdot 2 + \frac{1}{8} \cdot 3 + \frac{1}{8} \cdot 3 = \frac{7}{4}$$

$$L_\beta = \frac{1}{2} \cdot 2 + \frac{1}{4} \cdot 2 + \frac{1}{8} \cdot 2 + \frac{1}{8} \cdot 2 = 2$$

$$H = -\frac{1}{2} \cdot \log\frac{1}{2} - \frac{1}{4} \cdot \log\frac{1}{4} - 2 \cdot \frac{1}{8}\log\frac{1}{8} = \frac{1}{2} + \frac{1}{2} + \frac{3}{4} = \frac{7}{4}.$$

Note that the codification α is optimal because $L_\alpha = H$.

8.5.2 DECISION TREES

In this section, we present the concept of decision trees. Let us see one example. Imagine a game with two participants P and R. Player R chooses in secrecy an object O, and player P has to discover the object by asking questions to R, who can only answer YES or NO. Simplifying the problem, suppose there are four possible objects for R to choose, A, B, C, and, D, and that we know the probabilities that R chooses these objects,

$$p_A = \frac{1}{2}, p_B = \frac{1}{4}, p_C = \frac{1}{8}, p_D = \frac{1}{8}.$$

In Figure 8.4, we have two schemes for questions. The tree of Figure 8.4(a) indicates that initially we ask "Is the chosen object A?" If the answer is YES (represented by "1" in the figure), we know with certainty that the object is A. If the answer is NO ("0"), we ask "Is the object B?", and so on...

Notice that this problem of decision trees is equivalent to the problem of symbol coding, as the similarity of Figures 8.3 and 8.4 indicates.

The codification α in Table 8.1 in Section 8.5.1 is the codification of questions in the tree of Figure 8.4(a). For example, code "1" of symbol A in the table indicates the certainty of A if we have the positive answer for the first question. Code "01" of symbol B in table indicates

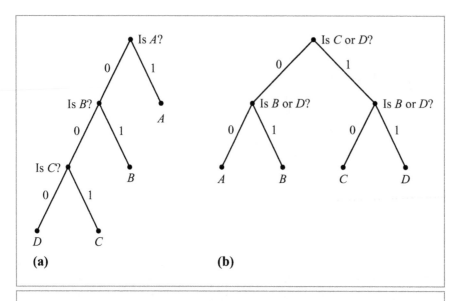

Figure 8.4: Trees for schemes (a) and (b).

the certainty of symbol B if the first answer is negative and the second is positive. The codification β in Table 8.1 is the codification of questions in the tree of Figure 8.4(b).

The results of Section 8.5.1 apply here and therefore the average number of questions cannot be less than the entropy. The scheme of questions in Figure 8.4(a) is optima because we have an average of $\frac{7}{4}$ questions, which is the value of the entropy. In the scheme of questions for Figure 8.4(b), we have an average of two questions to know the chosen object.

8.6 COMMENTS AND REFERENCES

The connections between probability and optimization are very rich and complex. In this chapter, we have explored mostly the relations involving information theory.

Another important link between the two areas lies in the use of optimization techniques to build statistical models. One example is the use of dynamic programming methods in the context of hidden Markov models (Rabiner, 1989). This topic is very relevant to machine learning and to many applications of computer vision.

BIBLIOGRAPHY

Eddington, S. A. S. *The Nature of the Physical World*. Maxmillan, MI: 1948.

Kapur, J. N., and H. K. Kesavan. *Entropy Optimization Principles with Applications*. Burlington, MA: Academic Press, 1992.

Klein, M. J. Thermodynamics in Einstein's Universe. *Science*, 509. Albert Einstein, quoted in the article, 1967.

Rabiner, L. R. A tutorial on hidden Markov models and selected applications in speech recognition. *Proceedings of the IEEE*, 77(2): 257–286, 1989.

Resnikoff, H. L. *The Illusion of Reality*. New York, NY: Springer-Verlag, 1989.

Shannon, C. E. A mathematical theory of communication. *The Bell System Technical Journal*, 27:379–423, 1948. *http://cm.bell-labs.com/cm/ms/what/shannonday/paper.html*.

Index

Printed and bound by CPI Group (UK) Ltd, Croydon, CR0 4YY

03/10/2024

01040329-0019